Integrated Vector Management

Integrated Vector Management

Controlling Vectors of Malaria and Other Insect Vector Borne Diseases

Graham Matthews

Emeritus Professor of Pest Management
International Pesticide Application
Research Centre
Imperial College, London, UK

WILEY-BLACKWELL

A John Wiley & Sons, Ltd., Publication

This edition first published 2011 © 2011 by John Wiley & Sons, Ltd.

Wiley-Blackwell is an imprint of John Wiley & Sons, formed by the merger of Wiley's global Scientific, Technical and Medical business with Blackwell Publishing.

Registered Office
John Wiley & Sons, Ltd, The Atrium, Southern Gate, Chichester, West Sussex, PO19 8SQ, UK

Editorial Offices
9600 Garsington Road, Oxford, OX4 2DQ, UK
The Atrium, Southern Gate, Chichester, West Sussex, PO19 8SQ, UK
2121 State Avenue, Ames, Iowa 50014-8300, USA
111 River Street, Hoboken, NJ 07030-5774, USA

For details of our global editorial offices, for customer services and for information about how to apply for permission to reuse the copyright material in this book please see our website at www.wiley.com/wiley-blackwell.

Library of Congress Cataloging-in-Publication Data

Matthews, G. A.
Integrated vector management : controlling vectors of malaria and other insect vector borne diseases / Graham Matthews.
    p.   cm.
  Includes bibliographical references and index.
  ISBN-13: 978-0-4706-5966-3 (hardcover : alk. paper)
  ISBN-10: 0-470-65966-1
1. Insects as carriers of disease–Integrated control.   2. Vector control.   3. Insect pests–Integrated control.   I. Title.   II. Title: Controlling vectors of malaria and other insect vector borne diseases.
  RA639.5.M38 2011
  614.4'32–dc23
                                                          2011018906

A catalogue record for this book is available from the British Library.

This book is published in the following electronic formats: ePDF [9781119950325]; ePub [9781119950332]; Mobi [9781119950349]

Set in 9/12.5pt Interstate Light by SPi Publisher Services, Pondicherry, India
Printed and bound in Malaysia by Vivar Printing Sdn Bhd

1 2011

Contents

Preface

Insects are responsible for the transmission of several extremely important diseases of man, including malaria, dengue hemorrhagic fever and onchocerciasis. While considering all major vectors, much of this book concentrates on the control of Anopheline mosquitoes, as the vectors of malaria. Globally, the number of cases of malaria is estimated at about 250 million with the greatest burden of this disease in sub-Saharan Africa, accounting for over 80% of the cases. Absenteeism from work causes an estimated reduction of economic growth of 1.3%. In 2006, WHO estimated that there were over 800,000 deaths, with 90% in the African region, where malaria is the cause of 17% of the mortality of children under 5 years of age.

One of the United Nations Millennium goals is to halt and begin to reverse the incidence of malaria and other major diseases by 2015. Much progress has been made and the Global Fund had approved funding for programmes that had distributed 104 million bed nets and treated 108 million cases of malaria by the end of 2009. However in Africa, where there is the greatest need to control malaria, work is hampered by poverty, weak health systems due to limited numbers of skilled health workers and problems of accessibility to those most in need.

A key method of reducing the deaths of children has been the advocacy of deploying insecticide treated bed nets to protect young children. Funding from key philanthropic, bilateral and multilateral sources and increased manufacture of these nets, from 30 million in 2004 to 95 million in 2007, has enabled net distribution in many countries, but there is still much progress still to be made to achieve the distribution of an estimated 250 million treated nets to reach 80% coverage in sub-Saharan Africa. As well as the deaths of young children, illness due to malaria affects vast numbers of people including farmers and others, thus affecting food security and economic productivity. If the control of vectors can be extended to become area-wide control programmes and encompass not just the control of mosquitoes, but also include vectors of other diseases, there is hope that productivity can be increased in sub-Saharan Africa and other less developed areas of the world so that poverty can be alleviated.

Controlling the vector by different techniques is increasingly important as the malarial parasite, *Plasmodium* spp. continues to develop resistance to drugs, thus effective reduction of the vector will reduce the number of people that will require treatment. Operational research to enable implementation or improvement of the tools already developed to control the insect vectors remains poorly funded. There is a need to develop practical combinations of vector control as part of integrated vector management programmes, yet already there is talk of eradicating malaria when much more research is needed to develop any new controls and as yet unproven methods for such a strategy to work.

Few trained in medicine will have any training in vector control, while specialist medical entomologists get little training in the use of insecticides. This book aims to provide an update on the methods of vector control that can be utilised today in the context of development since DDT was used against vectors. Future developments look beyond our present knowledge to assess whether any new approach, such as the deployment of genetically modified mosquitoes, can be added to our toolbox of control methods.

Acknowledgements

I wish to thank the following who have read and commented on one or more chapters and for supplying information: Andy Adams, Andy Bywater, Pierre Baleguel, Jane Bonds, John Clayton, Nigel Frazer-Evans, John Invest, Helen Pates Jamet, John Lucas, Bob Mickle, Peter Mukuka, Don Roberts, Graham White and especially Mark Latham, who has contributed significantly with his experience of mosquito control in the USA. Their contribution in their areas of specialist knowledge in vector control has greatly improved this book.

I am also indebted to the following for the supply of photographs and diagrams: Didier Baleguel, Roy Bateman, Clive Boase, Jane Bonds, John Clayton, Hans Dobson, Nigel Frazer-Evans, Chung Gait Fee, Ulrike Fillinger, Jörg Heckel, Eliningaya John Kweka, Mark Latham, Steve Lindsay, John Lucas, Bob Mickle, Peter Mukuka, Michael Reihle, Don Roberts, Werner Stahl, John Thomas. Figures 2.2 and 2.3 are reproduced from *The Excellent Powder – DDT's Political and Scientific History* by Donald Roberts, Richard Tren with Roger Bate and Jennifer Zambone, published by Dog Ear Publishing, Indianapolis, USA. Figure 4.1 is with permission of the Freer Gallery of Art, Smithsonian Institute, Washington DC: Purchase – Harold P. Stern Memorial Fund, F1995.17. I thank Robert Hudson for alerting me to the illustration shown in Figure 4.1

The work has been based on experience with vector control equipment at the International Pesticide Application Research Centre, Imperial College since 1972, participating with the World Health Organisation Pesticide Evaluation Scheme in relation to equipment for vector control and more recently with field work by the Yaounde Initiative Foundation, Cameroon. I thank Moira for her understanding and encouragement during my visits to Cameroon and the writing of this book.

Chapter 1
Introduction

Vector borne diseases, such as malaria, are responsible for 17% of the global burden of parasitic and infectious diseases. They result in avoidable ill-health and death, economic hardship for affected communities and are a serious impediment to economic development. Malaria causes over 800,000 deaths a year, 85% of which occur in children under 5 years of age. Many people, in 100 countries still affected by malaria transmission, are unable to work due to recurring attacks of malaria.

Malaria was endemic in parts of the USA only 75 years ago, but vector control alongside development programmes has eliminated the disease. However, mosquito control remains essential in the USA, not only because the biting insects are considered to be an unacceptable nuisance, but they can also transmit a number of arboviral encephalitides, including St. Louis, Eastern Equine, Western Equine, Lacrosse and West Nile viruses.

Malaria was also endemic in parts of Europe, where drainage schemes and application of Dichlorodiphenyltrichloroethane (DDT) during the 1940s led to its eradication, although still liable to return when travellers infected in the tropics return with the disease to areas where the vector species still occur. Sporadic occurrences of so-called 'airport malaria' happen every year in Europe or the USA, with a few locally transmitted cases reported, but with a suitable vector present such occurrences could become permanently established or re-established.

In addition to malaria, a number of other very important diseases are transmitted by insect vectors (Table 1.1). Some of the vector borne diseases are no longer a problem in many parts of the world with improved housing and living standards, but they have remained in remote rural areas and urban slums, and some have started to spread into other countries. The World Health Organisation (WHO) has grouped these as the Neglected Tropical Diseases (WHO, 2010a), as governments have not paid so much attention to these problems.

Aedes aegypti is an important vector of dengue viruses, especially in urban environments exposing over half of the world's population to the risk of

Table 1.1 Vector borne diseases (insects and arachnids).

1. Emerging diseases (adapted from Gratz, 1999)

Infection	Distribution	Vector*
Barmah Forest virus	Australia	Mosquitoes
Cat flea typhus	United States	Fleas
Cat-scratch disease	Global	Fleas
Dengue hemorrhagic fever	Americas, Asia	Mosquitoes
Human ehrlichiosis	Americas, Asia, Europe	Ticks
Kyasanur forest disease	India	Ticks
O'nyong-nyong fever	East Africa	Mosquitoes
Oriental spotted fever	Japan	Ticks
Oropouche virus	South America, Panama	Midges
Potasi virus	United States	Mosquitoes
Rocio virus	Brazil	Mosquitoes
West Nile Virus	United States, Europe	Mosquitoes

2. Resurgent diseases

Infection	Distribution	Vectors*
Chikungunya	Africa, Asia, Europe	Mosquitoes
Congo-Crimean Haemorrhagic fever	Africa, Asia, Europe	Ticks
Dengue	Africa, Americas, Asia	Mosquitoes
Filariasis-bancroftian	Africa, Americas, Asia	Mosquitoes
Japanese encephalitis	Asia	Mosquitoes
Leishmaniasis visceral	Africa, Americas, Asia	Sandflies
Leishmaniasis cutaneous	Global	Sandflies
Lyme disease	Global	Ticks
Malaria	Global	Mosquitoes
Onchocerciasis	Africa, South America	Blackflies
Plague	Africa, Americas, Asia	Fleas
Rift Valley fever	Africa	Mosquitoes
Ross River virus	Australia, Pacific islands	Mosquitoes
Trench fever	United States, Europe	Body lice
Trypanosomiasis	Africa	Tsetse flies
	South America	Triatomid bugs
Venezuelan equine encephalitis	Americas	Mosquitoes
Yellow fever	Africa, Americas	Mosquitoes

* Some information on the vectors is given in text, but this book is primarily aimed at control methods so for further information on the taxonomy of the vector species and their biology readers should consult Becker et al. (2010) and Marquardt (2004).

infection. Other dengue vectors are *Ae. albopictus* and *Ae. polynesiensis*. Annually there are an estimated 50-100 million cases of dengue fever (DF), and 250,000-500,000 cases of dengue haemorrhagic fever (DHF) throughout the world. An outbreak of West Nile fever in Greece in 2010, with over 230 cases and 30 deaths, has highlighted the importance of re-establishing vector control programmes in areas that dropped them after the eradication of malaria over 40 years ago. Recently, dengue has been reported in France. Environmental management and insecticide application are the main methods of controlling the vector to reduce dengue.

With the speed and volume of international travel having increased so dramatically over the past quarter century, it is not surprising that the emergence and re-emergence of arthropod borne diseases in developed countries is becoming commonplace. The introduction of West Nile virus to the USA in 1999 is a prime example of this. Since arriving in New York City, there have been over 30,000 human clinical cases and over 1,000 deaths. At its peak in 2003, there were almost 10,000 cases, although this has settled down to an average of 1,000 cases per year over the past couple of years.

The currently ongoing dengue outbreak in Key West, Florida (60 locally acquired cases in 2010) that started in 2009 is an example of the re-emergence and re-establishment of vector borne diseases in areas where they had previously been eradicated. Apart from sporadic local transmission along the US-Mexican border, there had been no local epidemics of dengue in the USA in 50 years, and none in Florida since 1934.

Certain insects that spread diarrhoeal diseases and trachoma mechanically are also included. The principal methods of applying insecticides to control these vectors involve manually operated, motorised and vehicle mounted equipment and aerial applications (Table 1.2). Further details of the methods of application are given in *Pesticide Application Methods* (Matthews, 2000), but sufficient detail is included with the different approaches to control discussed in the following chapters, which will enable those involved in vector control programmes to choose and understand the different application techniques that are needed.

One of the techniques may dominant in certain situations, but more than one approach may be needed to have a real impact on vector populations within an area. Large-scale vector control programmes are essential if the incidence of disease is to be lowered, thus reducing the number of patients who need to be treated and reducing the demand for the drug treatment, which should delay the selection of resistance to the drug. Improvement in rapid diagnosis of malaria is another factor to ensure drug treatments are directed at those infected with the parasite.

Using different techniques as part of integrated vector management (IVM) is discussed in Chapter 6. A key question is inevitably the cost of adopting more than one control technique, but clearly no one method is going to be effective under all situations. Indoor residual spraying (IRS, Chapter 2) protects people from bites from vector species that enter houses, but will not protect people working in the open. Space treatments (Chapter 3) generally have no residual activity and may need to be repeated several times sequentially to have a significant effect on the vector population, but can be important, especially to rapidly reduce a vector population during a disease epidemic. Area-wide space treatments may also reduce vector populations sufficiently at a critical period, to minimise disease transmission over an

Table 1.2 Main methods of applying insecticides for vector control.

Vector	Control technique	Main method/Equipment
Mosquitoes		
Anopheles spp.	Indoor residual spraying	Compression sprayer
	Space treatment	Cold or thermal fog
	Barrier	Mist applicator
	Larviciding	Granule application
	Bednets	Impregnated net
Aedes aegypti and *Aedes albopictus*	Space treatment	Fogging equipment
	Larviciding	Granule application
Aedes simpsoni	Space treatment	
Other *Aedes* spp. and *Psorophora* spp.	As *Anopheles* except indoor residual spraying is NOT used	
Culex spp.	As *Anopheles* except indoor residual spraying is of limited use	
Mansonia spp.	Indoor residual spraying	
Flies		
Musca spp, *Stomoxys* spp. *Calliphora* spp. etc.	Residual sprays	Small equipment, Compression sprayers, Power sprayers
	Space treatment	Fogging
	Baits	Granule application, impregnated tapes
	Traps	
Glossina spp.	Space treatment	Aerial spray application
	Screens	Impregnated, baited screen
Simulium damnosum complex	Larviciding	Boat or aerial application
Phlebotomus spp.	Bed nets	Treated mesh
	Indoor residual spraying	Compression sprayer
Triatomine Bugs		
Triatoma, Panstrongylus and *Rhodnius* spp.	Indoor residual spraying	Compression sprayer
Bedbugs		
Cimex spp.	Residual sprays	Small sprayers, Compression sprayers
	Bed nets	Treated mesh
Fleas		
Xenopsylla spp. and *Pulex* spp.	Residual sprays	Small sprayers Compression sprayers
Ticks		
Ixodes spp. *Rhipicephalus* sp. etc.	Residual sprays	Small sprayers Compression sprayers

extended period. Insecticide treated bed nets (Chapter 4) undoubtedly protect those who sleep under the net, but they may have to get up during the night and go outside, so may still be bitten by a vector. For some vector species, larviciding (Chapter 5) is the most effective control technique. In each case, management of an area-wide programme is crucially important to have a significant impact on disease transmission.

Insect vectors

Apart from the importance of knowing which vector species are present in an area, and this may vary over relatively short distances, depending on the availability of suitable breeding sites, a crucial need in using insecticides is to determine the susceptibility of the vector species to insecticides, so that resistance can be detected, if present, and the control programme adjusted as necessary.

Distribution of vectors

The main vectors considered in this book are problems of poverty, where poor housing and lack of medical services have resulted in the prevalence of malaria, onchocerciasis, Chagas disease and DF. The global distribution of malaria is shown in Figure 1.1. The distribution of vectors within each area is important with endophilic and exophilic biting and different larval breeding sites, ranging from small containers to large ponds and some sections of rivers. Progress on elimination of malaria, namely the interruption of transmission at a national or regional level, has been made in some countries. The current strategy is 'to shrink the malaria map' further by attempting to eliminate malaria from the countries at the periphery of the main area affected, with 32 countries at present pursuing an elimination strategy (Feachem et al., 2010). These countries generally have a single period each year, related to their rainfall distribution, when mosquitoes transmit malaria. The greatest problems will be in the countries with rainfall during most months throughout the year.

In all the countries with malaria, whether aiming at elimination or reducing endemic malaria transmission to a sufficiently low level that it does not cause a major public health burden, vector control is crucial. In each country, surveillance of malaria must be on-going and include determination of the distribution of vectors within each area. Variation between different vector species with endophilic and exophilic biting, different larval breeding sites, and other behavioural patterns, needs to be known so that control programmes can be adapted, especially when transmission is reduced to a low level.

Mosquitoes

Mosquitoes are vectors of malaria, filariasis and dengue, as well as yellow fever, West Nile fever and other encephalities. Mosquitoes have an aquatic larval stage (Figure 1.2), the location of which will vary between species, so

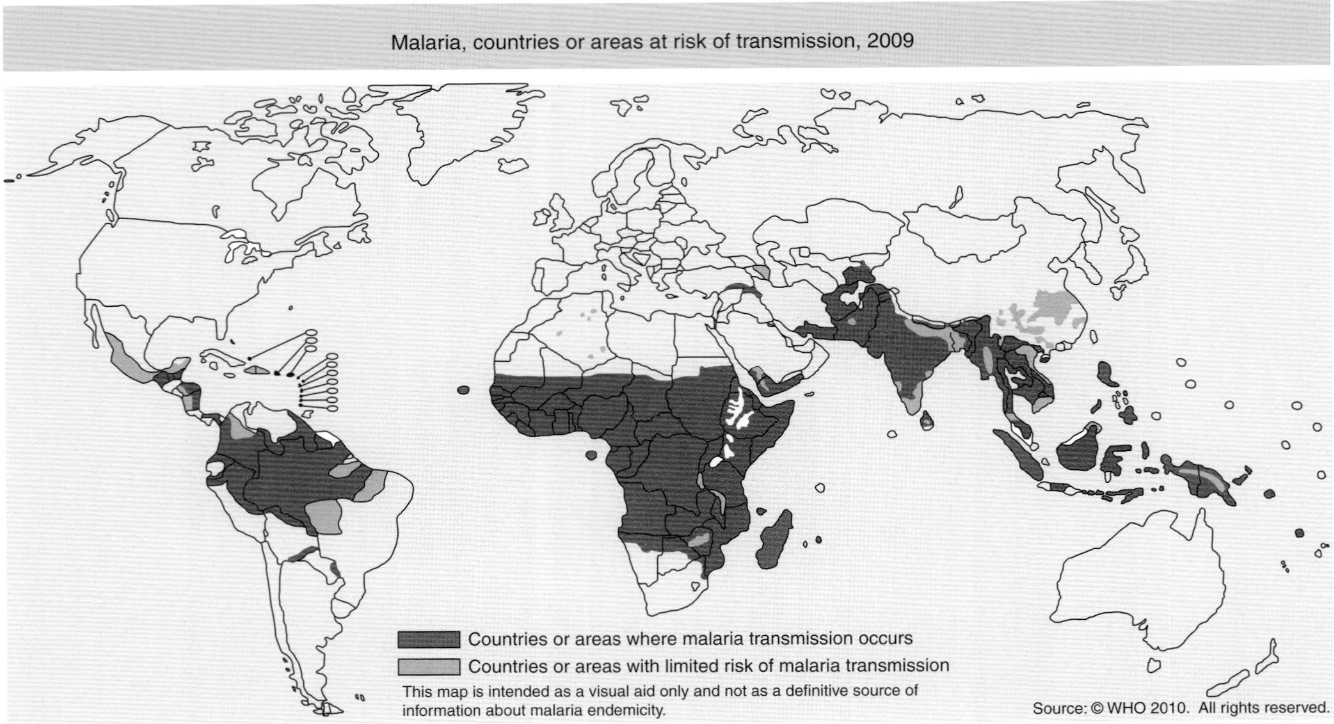
Malaria, countries or areas at risk of transmission, 2009
Countries or areas where malaria transmission occurs
Countries or areas with limited risk of malaria transmission
This map is intended as a visual aid only and not as a definitive source of information about malaria endemicity.
Source: © WHO 2010. All rights reserved.

The boundaries and names shown and the designations used on this map do not imply the expression of any opinion whatsoever on the part of the World Health Organization concerning the legal status of any country, territory, city or area or of its authorities, or concerning the delimitation of its frontiers or boundaries. Dotted lines on maps represent approximate border lines for which there may not yet be full agreement.

Data Source: World Health Organization
Map Production: Public Health Information and Geographic Information Systems (GIS)
World Health Organization

World Health Organization

Figure 1.1 Distribution of malaria.

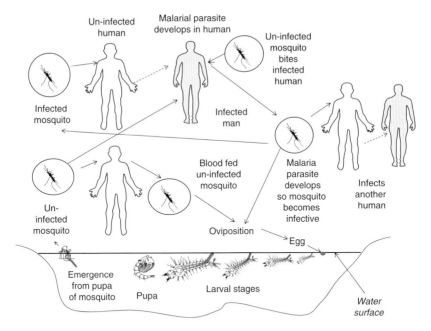

Figure 1.2 *Anopheles* spp. Life-cycle (malarial parasite develops in man and the adult mosquito).

removal of breeding sites, where possible, is a key management tool, especially close to dwellings. The severity of mosquito borne diseases is often a function of the amount and extent of rainfall within a region affecting the area in which the larvae can breed. Management of aquatic environments with predatory fish and reduction of areas suitable for larvae is required to reduce breeding in ponds, canals and other areas of permanent water. As the female mosquitoes need to have a blood meal, preventing access to dwellings by having netting across windows and other openings is also crucial for endophilic species, but chemical control is still the main tool to reduce mosquito populations and thus the transmission of disease. Chemical control is also used extensively in areas where the mosquitoes are considered to be a major nuisance due to their bites (Figure 1.3), even though they may not be a disease vector.

Anopheles spp.

Vectors of malaria are certain species of the genus *Anopheles*. A single bite from an infected female *Anopheles* mosquito can transmit the disease, with children under 5 years of age being the most susceptible (Figure 1.4). The main activity of the female *Anopheles* is during the night, when the species that enter houses (endophilic) can be controlled effectively by treating inside walls on which they may rest. IRS can kill a mosquito any time it enters a house for a blood meal, which it needs to have every 2–3 days, thus few will survive long enough (~12 days) for the malaria parasites to complete the period of their life-cycle in the

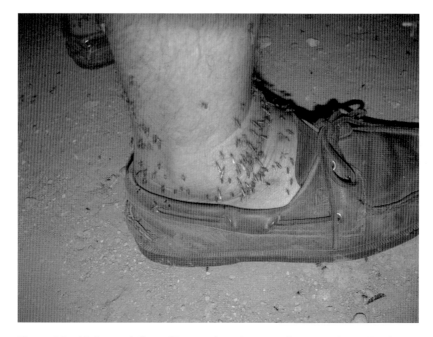

Figure 1.3   High population of hungry female mosquitoes seeking a blood meal (photo: Mark Latham).

Figure 1.4   Baby unprotected by a net.

vector mosquito. Some species are more exophilic, i.e. they can bite people outside dwellings. Where these species are more important, outdoor space or barrier treatments near dwellings may be required. In areas where people are living in temporary accommodation such as tents, the vectors need to be controlled by spraying the tents and using insecticide treated bed nets.

Aedes spp.

Two of the most important species in the Aedes genera are *Ae. aegypti* and *Ae. albopictus,* as they are the vectors of dengue and yellow fever, although in some areas there are related species that transmit arboviruses and filariasis. The larval stage of dengue vectors can develop in quite small volumes of water in various man-made containers, such as discarded bottles, vases and stacks of tyres, although *Ae. albopictus* can also use natural breeding sites including tree holes, plant axils and discarded parts of tropical fruit such as coconut husks. *Ae. aegypti* is a major problem in urban areas where there are many suitable breeding sites, especially in areas where water supplies are kept in containers without lids (Phanthumachinda, 1978). They will enter dwellings and rest on inside walls, but the main emphasis has been on larval control primarily through premise sanitation, supplemented when populations are extremely high, by space treatments. However, larval control is not always effective (Phuanukoonnon et al., 2005), as large containers of water for bathing and flushing toilets frequently contain larvae. Keeping fish and covering jars of drinking water are most effective, while a larvicide is used most where the insecticide is readily available in small packets. The study illustrated the need to understand the way in which people store water to devise effective methods of control. Sleeping under a bed net is also advised, especially for babies and to protect a person diagnosed to have DF, to minimise transmission of the virus to others. Guidelines on dengue protection and control have been published (WHO, 2009). However, the two vector species bite at different times, one primarily early morning after dawn and the other during late afternoon, so bed nets are not so appropriate as their use in malaria control, as *Anopheles* spp bite later in the evening and during the night.

Culex spp.

Some *Culex* spp. bite late at night and can transmit diseases such as encephalitis, but they are more a nuisance pest than a vector. They normally rest outdoors, but some will enter houses, and tend to rest on curtains and furniture rather than on wall surfaces, so IRS is not recommended. *Cx. quinquefasciatus* can breed in water in latrines, drains and septic tanks. Drains need to be kept unblocked to maintain water flow, while free water in pit latrines can be covered by a layer of expanded polystyrene beads, 1cm deep, to deter breeding. Curtis et al. (2002) concluded that where *Cx. quinquefasciatus* was breeding mainly in pits, polystyrene-bead layers could assist considerably in the process of eliminating lymphatic filariasis in conjunction with mass drug administration.

Figure 1.5 Biconical trap for sampling tsetse flies in West Africa.

Flies

Simulium spp.

Female black flies of *Simulium damnosum* complex in Africa and other species in Central and South America transmit the parasite that causes river blindness. Extensive control using larvicides has been carried out in the savannah areas of West Africa over a 20-year period to interrupt transmission of the disease. However, *Simulium* remains a major vector in other areas and, in addition to being a disease vector, causes extreme irritation by constantly biting fishermen and others working close to the major rivers and their tributaries. The problem seems to have increased by the presence of hydro-electric dams, due to spillways creating the ideal habitat for black fly larvae. The impact on the local economy is highly significant, as people prefer to migrate to areas without the fly. Application of larvicides is the only recommended technique, although this is not easy where rapids on small rivers are not accessible. Further work is also needed to determine if those working in the field can be better protected.

Glossina spp. tsetse flies

Several species of tsetse flies are vectors of human and animal trypanosomiasis in Africa. Sleeping sickness is a chronic infection caused by *Trypanosoma brucei gambiense* (*T.b.g.*) in West and Central Africa and *Trypanosoma brucei rhodesiense* in east and southern Africa. According to WHO, the number of cases of sleeping sickness reported in 2009 was below 10,000 for the first time in 50 years, due to continued vector control efforts. Populations of tsetse flies have been sampled using various designs of traps (Figures 1.5 and 1.6), with the bi-conical trap used mostly in detecting riverine species where human trypanosomiasis occurs.

Figure 1.6   Tsetse trap in Zimbabwe.

Large-scale control of the vector can be carried out by sequential aerially applied space treatments timed to affect newly emerging adult flies (Allsopp, 1990; Kgori et al., 2006). In some locations, the use of thermal fogging has also been used. These space treatments are discussed in Chapter 3. However, in some countries, an alternative method has been to use screens treated with insecticide, usually a pyrethroid, and baited with octanol to attract the flies to rest on the screen. The difficulty with the screens, usually referred to as targets, is maintaining them in the field, especially when rural roads are made inaccessible by heavy rainfall (Figures 1.7 and 1.8).

In Kenya, electrocuting sampling devices were used to show that nearly twice the number of *Glossina fuscipes fuscipes* (Newstead) visited a biconical trap compared to a black target (100 cm × 100 cm), but only 40% of the males and 21% of the females entered the trap, whereas 71% and 34%, respectively, alighted on the target. According to Lindh et al. (2009), the greater number visiting the trap appeared to be due to its being largely blue, rather than being three-dimensional or raised above the ground. The study showed that a blue-and-black panel of cloth (0.06 m²) flanked by a panel (0.06 m²) of fine black netting, placed at ground level, would be about ten times more cost-effective than traps or large targets in control campaigns. Clearly this will be important in relation to controlling all subspecies of *G. fuscipes*, which are currently responsible for more than 90% of sleeping sickness cases.

In trials in West Africa, as only 50% of the flies attracted to the vicinity of the trap were actually caught by it, an analysis of the visual responses and identification of any semio-chemicals involved in short-range interaction might lead to improved trap efficiency (Rayaisse et al., 2010).

Figure 1.7 Insecticide treated screen to attract and kill tsetse flies in Zimbabwe.

Figure 1.8 Re-treating a tsetse screen in the field in Zimbabwe.

For trypanosomiasis in cattle, insecticides have been applied directly to the cattle, using a 'pour-on' formulation that is spread around the animal within the surface waxes in the skin. Whatever techniques are adopted, the crucial need is for large-scale operations in relation to the extent of the area populated by tsetse flies.

In Uganda, farmers have been encouraged to have the undersides of cattle (belly and legs) sprayed with a knapsack or similar sprayer to reduce the local tsetse fly population (Torr et al., 2007) and thus reduce the incidence of sleeping sickness, as part of a public private partnership "Stamp Out Sleeping Sickness" (SOS) campaign.

Phlebotomine sand flies

Only two genera of sand files are vectors of leishmaniasis disease, namely *Phlebotomus* of the Old World and *Lutzomyia* of the New World. The extremely small Phlebotomine sand flies seldom exceed 3 mm in length and are crepuscular or nocturnal, although a few species will bite during daylight. They rest in cool and humid areas of houses, including latrines as well as animal houses, and in natural habitats, such as caves, walls, birds' nests and in burrows, where they come into contact with rodents and other mammals. Most species are exophilic, but some are endophilic and will bite indoors.

Leishmaniases have increased in importance in areas with deforestation, new dams and irrigation schemes, urbanisation and migration of non-immune people to endemic areas, affecting socio-economic progress, especially in the Amazon Basin, the tropical regions of the Andean countries, Morocco and Saudi Arabia. IRS is only suitable for endophilic species, but may not be cost effective where disease prevalence is low. Improving houses by plastering walls is also effective in reducing the sand fly population.

The use of insecticide treated or untreated bed nets has also been suggested, but as the peak of biting activity of most vector species is shortly after sundown before children go to bed, it may not have a dramatic impact (Killick-Kendrick, 1999). Recently, the use of fine mesh nets to reduce the number of sand flies entering an enclosed area has been suggested (Faiman et al., 2009).

Military personnel in the Middle East and Afghanistan are at risk from leishmaniasis and have been advised to wear permethrin treated uniforms, apply DEET repellent to exposed skin and use permethrin-treated bed nets to prevent sand fly bites (Aronson, 2007).

Musca domestica and other *synanthropic* spp.

The house fly, *Musca domestica*, is the most common species worldwide, which with other flies – biting flies (*Stomoxys* spp.), blowflies (*Calliphora* spp., *Lucilia* spp.) and flesh flies (*Sarcophaga* spp.) – is very important in spreading diseases, such as shigellosis and other diarrhoeal diseases. In Africa, trachoma, caused by *Chlamydia trachomatis*, is spread by the Bazaar fly (*Musca sorbens*), which is attracted to human eyes (Emerson et al., 1999). Good sanitation is essential to reduce fly populations, with proper disposal of refuse and any organic waste, as well as screening of houses, especially in areas where food is prepared. Chemical control should be limited to spot treatment, except when large numbers of flies are present on refuse dumps when a space treatment may be required.

Other vectors

Triatomine bugs

Several species of *Triatoma*, *Panstrongylus* and *Rhodnius* are vectors of *Trypanosoma cruzi*, which causes Chagas disease in Central and South America. The most important vector in most of South America is *T. infestans*, while *R. prolixus* is important in Central America and the northern part of South America. These bugs hide in cracks and similar secluded places in walls and ceilings of human dwellings and animal houses. Apart from IRS, the main control is by house improvement. In Argentina, fumigant canisters, including beta-cyfluthrin, were designed for surveillance, but have been used in community based control programmes. Slow release paints containing malathion have also been used in Brazil. WHO has published a guide for field testing and evaluation of insecticides for IRS against Chagas vectors (WHO, 2001).

Several other insects, including cockroaches, which can spread disease, are discussed in Chapter 7.

Chemical control

Chemical control is an important intervention technique to reduce populations of vector species and thus reduce their spread of disease. Concern about exposing people to insecticides necessitates minimising their use and limiting the choice of insecticide to those which are the least toxic to humans. Procedures for safe use during the delivery, storage and use of pesticides are a crucial part of the life-cycle approach to public health pesticides used in vector control programmes. Registration of pesticides has traditionally been done by a national organisation and has been primarily concerned with those used in agriculture. However, in Europe, primary approval of the active ingredient in pesticide formulations is carried out by the European Union (EU), with member countries concerned with the products marketed in their country. Greater harmonisation of the data requirements for registration and closer collaboration at a regional level is important, as many countries lack sufficient trained scientists to assess all the data needed for registration. Trained staff are needed at all stages in the management of public health pesticides, from registration, procurement, storage, use and disposal of waste (Figure 1.9). WHO has published generic risk assessments for the main methods of using pesticides, so that those involved in the planning and execution of chemical control programmes can make appropriate decisions to minimise exposure to spray teams, persons living in treated houses and bystanders.

Hazard and toxicity

Many of the insecticides used to kill insect vectors have a broad spectrum and thus are potentially harmful to other 'non-target' organisms, including humans.

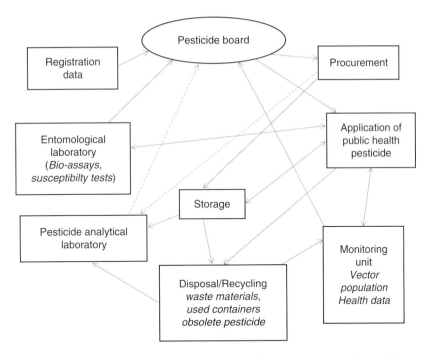

Figure 1.9 Diagram to show stages in registration and use of public health pesticides. Ideally, a pesticide board will collate data, including the local susceptibility and monitoring data, register pesticides and issue instructions on procurement and their use. The diagram shows the need for procurement officials to seek confirmation of quality. Similar correct storage is important with checks on quality and disposal of waste.

In assessing whether those involved in its use are likely to suffer harm, a judgement is made of the risk that exposure to it will cause a harmful effect (WHO, 1999). The assessment needs to take account of the dosage at which individual insecticides are applied. For vector control, dosages used in space sprays are a small fraction of those needed for residual treatments. Reading the product label to ensure the correct dose is applied in relation to the method of application is crucial.

Toxicity

The toxicity of insecticides to humans is assessed initially as the acute oral toxicity to rats to determine the dose measured in milligrams of active ingredient per kilogram body weight that would kill 50% of a population of test animals, known as the LD_{50}. Since 1975, WHO has published a list of pesticides with data based on acute toxicity (WHO, 2010b). These measurements provide a basis on which different insecticides can be compared, but the data is not indicative of the actual hazards involved in using the insecticide. Exposure to the insecticides will vary, depending on tasks being carried out and the method of application used. The persons who prepare the sprays and apply them inside houses need to be protected, so that exposure to the spray is avoided as far as possible by wearing protective clothing. Their exposure will be different to those who can be inside a vehicle cab while applying an outdoor space spray.

Similarly, exposure of the occupants of houses will be quite different, as they are mainly concerned with dermal exposure when touching treated surfaces.

The duration of exposure will vary among these people, while uptake will be affected by the way the insecticide is formulated. Assessments of exposure to other non-target species are now needed as part of an environmental assessment of the use of the insecticides. These are particularly important where larvicides are applied directly to water. A major concern is if pesticides should get into water supplies. The EU imposes a maximum of 0.1µg/l for a single pesticide and if more than one pesticide is present, the maximum total amount permitted is 0.5µg/l in water extracted for drinking. In the USA under the Clean Water Act, those involved in mosquito control required a permit under the National Pollutant Discharge Elimination System after April 2011 and have to follow strict administrative and application guidelines to control vectors, although proposals are being made to obtain an exemption for registered public health pesticides.

Individual generic guidelines on risk assessment published by WHO cover IRS, space treatments, use of insecticide treated bed nets and larviciding. In these, estimates of exposure are compared with acceptable exposure levels relevant to the different application methods.

Acute and chronic toxicity data is considered when registering a pesticide, but the impact of long-term exposure is not measured, as there will be significant differences in exposure. The extent to which individuals are exposed to low doses will depend on the activities of the people concerned and the dose used in a particular treatment. Furthermore, the body can metabolise many of the insecticides that get into it and then excrete the metabolites. Even the persistent insecticide DDT, which can be stored in body fat, has not caused any mortality, even when people have been exposed over long periods. The main concern with DDT is that primiparous mothers from areas with malaria, where houses had been sprayed, recorded significantly more DDT and its metabolites in breast milk than multiparous mothers in the same area, and significantly much higher than mothers in a control area not exposed to DDT (Bouwman et al., 1990). Nevertheless, the benefits of DDT in controlling malaria outweigh some disadvantages. However, when considering the costs of using insecticides, the management of an insecticide throughout its whole life-cycle now needs to be taken into account, so when an insecticide, such as DDT, requires a higher dosage, the higher dose per unit area can also affect transport costs and the costs of any waste disposal that may be needed.

Pyrethroids are now extensively used as they are considered to be less hazardous to use at the very low dosages required and because of their low mammalian toxicity. Care is needed with some pyrethroids, which can cause a peripheral skin irritation, but the risk of this is reduced with encapsulated suspension formulations and proper use of protective clothing.

Table 1.3 Examples of insecticides used in vector control.

Group	Examples of insecticides used in vector control	Acute oral toxicity (LD$_{50}$) to rats (mg/kg)*
Organochlorine	DDT	118
Organophosphate	Malathion	1,375–2,800
	Temephos	4,204–>10,000
	Fenitrothion	1,700
	Pirimiphos methyl	1,414
	Chlorpyrifos methyl	>3,000
	Naled	430
Carbamate	Bendiocarb	40–156
	Propuxor	50
Pyrethroid	$\tilde{\alpha}$-cypermethrin	79–400
	Bifenthrin	54
	Bioresmethrin	7,000–8,000
	Cyfluthrin	500
	Cyphenothrin	318–419
	Deltamethrin	135–>5,000
	D-Phenothrin	>5,000
	Etofenprox	>42,880
	$\tilde{\lambda}$-cyhalothrin	56–79
	Metofluthrin	>2,000
	Permethrin	430–4,000
	Resmethrin	>2,500
Chitin inhibitor	Diflubenzuron	>4,640
	Teflubenzuron	>5,000
	Triflumuron	>5,000
Suppresses embryogenesis	Pyriproxyfen	>5,000
Bio-pesticide	*Bacillus thuringiensis*	
	B. sphericus	

* In some cases, the toxicity will depend on the isomer of the chemical and whether the test was on male or female rats.

Insecticides

Insecticides have not been developed specifically for vector control, so those now recommended for vector control (Table 1.3) were marketed initially for use in agriculture. A few active against mosquitoes and other vectors and with low mammalian toxicity have been selected for use in vector control. Each country or region should have legislation to cover all aspects of the use of pesticides, not only for use in agriculture, but also for use as public health pesticides. The choice of which insecticides may be used in vector control should be decided in each country or region by an independent scientific committee, to take account of data on resistance and other factors, so that their use is co-ordinated. WHO has published guidelines on public health pesticide management policy for the WHO African region (WHO, 2011).

Historically, the natural pyrethrins have been used for centuries as dried flowers of *Chrysanthemum cinerea* and more recently as formulated sprays, but its short persistence stimulated the research that led to the photostable pyrethroid insecticides, which have become the dominant class of pesticide recommended for vector control. The alternatives are mainly organophosphates and carbamate insecticides. Organochlorines are now regarded as too persistent and their use has been banned in agriculture, but as DDT was so effective against mosquitoes, its use is still conditionally permitted under the Stockholm Convention for Persistent Organic Pollutants for IRS in malaria control, where the local vector population is not resistant to it.

Insect growth regulators (IGR), including juvenile hormone analogues, such as methoprene and pyriproxyfen, and chitin synthesis inhibitors, such as diflubenzuron as well as biopesticides, using *Bacillus thuringiensis israelensis* (Serotype H-14) and *B. sphaericus*, are used against immature stages of vectors. Thus, there is a wider selection of insecticides for use as larvicides than is currently available for application as adulticides.

Detailed information on individual insecticides can be downloaded from the web pages of the Environmental Protection Agency (EPA) in the USA and from other regulatory authorities. The common name of individual pesticides is used in this book. Commercial companies use different trade names, depending on whether their product is intended for agriculture or use as a public health insecticide. Information on these is available at the company's web page.

As continuous use of an insecticide with a particular mode of action inevitably leads to selection of a population resistant to the insecticide, there is a major concern that vector control urgently requires the availability of new insecticides that have a different mode of action. Unfortunately new developments in agriculture have been towards chemicals that move systemically in plants and mainly control pests such as aphids and whiteflies, with poor activity against dipterous pests. With this lack of new insecticides, the Liverpool School of Tropical Medicine has a research programme on insecticide resistance from molecular characterisation to development of practical interventions for resistance management. With finance from the Gates Foundation, the Innovative Vector Control Consortium (IVCC) has been established as a Product Development Partnership (PDP), which in collaboration with industry is searching for potential new insecticides for public health vector control and aims to develop information systems and tools, such as insecticide quantification kits, which will enable new and existing pesticides to be used more effectively.

WHO recommendations

WHO set up a pesticide evaluation programme (WHOPES), in which manufacturers can submit samples of their products, including insecticide treated bed nets, for independent testing for use in vector control. Laboratory tests conducted at WHO Collaborative Centres are followed by field trials.

Results of these tests are published on the WHO web page and in reports of WHOPES Working Group Meetings. Such information is intended to help national authorities in the registration of pesticides. Specific recommendations for different vector species are published in *Pesticides and their Application* (WHO, 2006). Recommendations related to different application techniques are discussed in the following chapters.

Formulations

The active material or technical material has to be mixed with various other ingredients that have no toxic effect, to make a formulation that is easy to use. An exception has been the application of technical malathion, an organophosphate insecticide that can be sprayed as a liquid at ultra-low volume rates. Similarly, naled has also been used in the USA. The proportion of active ingredient in a formulation varies, but is generally no more than 10% with the most active insecticides. Different formulations are required, depending on the type of application. Industry identifies different formulations with a two-letter code (Table 1.4).

Spraying absorbent surfaces such as mud walls requires a particulate formulation, so that the particles remain on the surface and are not absorbed into porous surfaces. Suitable formulations are suspension concentrates, micro-encapsulated suspensions, wettable powders and dispersible granules.

Ultra-low volume cold fogs require a low volatile formulation, so that droplets do not shrink in diameter in flight. Many ultra low volume (ULV) formulations are now designed for dilution in water (Groome et al., 1989). Formulations designed to be applied as a thermal fog can be diluted easily in kerosene or diesel.

In emulsifiable concentrate formulations, the active ingredient is mixed with a solvent and an emulsifier so that the mixture readily forms an emulsion when mixed with water. The solvents are usually very volatile so they are not suitable for space treatments. They should not be used on porous surfaces, as the spray deposit is readily absorbed and no longer active on the surface. The presence of the solvent also increases the absorption of spray through the skin and is thus more hazardous to use.

In some situations, dusts, granules or larger briquettes are required. These are formulated such that no mixing is required by the user. The small particle size in dusts increases the risk of inhalation, so use of dusts has declined and is now limited to small-scale use, mainly against lice and fleas. Granules are larger so there is no risk of inhalation. They are used mainly with larvicides.

There are a number of other special formulations, including tablets, gels and baits. Tablets are essentially a large format of the dispersible granule and contain a chemical that effervesces on contact with water to aid dispersion in the spray tank. Gels are a viscous paste-type formulation used mostly in cockroach control, as the gel can be applied easily to specific sites. Baits are when a food additive is added to the insecticide to increase its consumption and improve activity in the insect stomach.

Table 1.4 Codes for different insecticide formulations (extracted from list on the Crop Life International web page).

Code	Term	Definition
AE	Aerosol dispenser	A container-held formulation, which is dispersed generally by a propellant as fine droplets or particles upon the actuation of a valve
BR	Briquette	Solid block designed for controlled release of active ingredient into water
DT	Tablet for direct application	Formulation in the form of tablets to be applied individually and directly in the field, and/or bodies of water, without preparation of a spraying solution or dispersion
CS	Capsule suspension	A stable suspension of capsules in a fluid, normally intended for dilution with water before use
DP	Dustable powder	A free-flowing powder suitable for dusting
EC	Emulsifiable concentrate	A liquid, homogeneous formulation to be applied as an emulsion after dilution in water
EW	Oil in water emulsion	A fluid, heterogeneous formulation consisting of a solution of pesticide in an organic liquid dispersed as fine globules in a continuous water phase
GL	Emulsifiable gel	A gelatinised formulation to be applied as an emulsion in water
GR	Granule	A free-flowing solid formulation of a defined granule size range ready for use
HN	Hot fogging concentrate	A formulation suitable for application by hot fogging equipment, either directly or after dilution
KN	Cold fogging concentrate	A formulation suitable for application by cold fogging equipment, either directly or after dilution
LN	Long-lasting insecticidal net	A slow- or controlled-release formulation in the form of netting, providing physical and chemical barriers to insects. LN refers to both bulk netting and ready-to-use products, for example mosquito nets
MC	Mosquito coil	A coil which burns (smoulders) without producing a flame and releases the active ingredient into the local atmosphere as a vapour or smoke
RB	Bait (ready for use)	A formulation designed to attract and be eaten by the target pests

Table 1.4 (*Continued*)

Code	Term	Definition
SB	Soluble bag	'SB' should be added to the formulation code, if the material is packaged in a sealed water soluble bag (e.g. WP-SB)
SC	Suspension concentrate (= flowable concentrate)	A stable suspension of active ingredient(s) with water as the fluid, intended for dilution with water before use
SE	Suspo-emulsion	A fluid, heterogeneous formulation consisting of a stable dispersion of active ingredient(s) in the form of solid particles and of water-non-miscible fine globules in a continuous water phase
SG	Water soluble granule	A formulation consisting of granules to be applied as a true solution of the active ingredient after dissolution in water, but which may contain insoluble inert ingredients
SO	Spreading oil	Formulation designed to form a surface layer on application to water
SP	Water soluble powder	A powder formulation to be applied as a true solution of the active ingredient after dissolution in water, but which may contain insoluble inert ingredients
SU	Ultra-low volume (ULV) suspension	A suspension ready for use through ULV equipment
TC	Technical material	A material resulting from a manufacturing process comprising the active ingredient, together with associated impurities. This may contain small amounts of necessary additives
UL	Ultra-low volume (ULV) liquid	A homogeneous liquid ready for use through ULV equipment
VP	Vapour releasing product	A formulation containing one or more volatile active ingredients, the vapours of which are released into the air. Evaporation rate is normally controlled by using suitable formulations and/or dispensers
WG	Water dispersible granules	A formulation consisting of granules to be applied after disintegration and dispersion in water
WP	Wettable powder	A powder formulation to be applied as a suspension after dispersion in water
WT	Water dispersible tablet	Formulation in the form of tablets to be used individually, to form a dispersion of the active ingredient after disintegration in water

Packaging and storage

For long distance transportation, manufacturers prefer large containers, such as 200 litre drums, but end users need more specific small containers related to the way the insecticide is used. With portable spraying equipment with usually less than 10-15 litre capacity, formulations are ideally packaged in sachets, so that one sachet per sprayer load provides the correct concentration of spray. Their use significantly reduces exposure of the spray operator to the insecticide. Some sachets are made with a water soluble plastic, so the whole sachet can be dropped into the spray tank. Prior to use, such sachets must be kept dry and may be packaged in an outer foil wrapping.

Where larger containers are supplied, the user will require an appropriate graduated cup or jug to measure the correct quantity to use in a small sprayer. A funnel or bucket, incorporating a filter, is often needed when transferring insecticide from its original container or diluted spray to the sprayer tank.

Prior to use, insecticides need to be stored in a well ventilated secure building, which is protected from rain, direct sunlight and flooding. Children and other unauthorised people should not be allowed to enter the stores. Small stores need to be located close to the spray operations, so that unused insecticide can be easily returned to the store. Larger stores can accommodate the bulk of the insecticide further away, provided roads are accessible to permit delivery of further supplies when needed during a control campaign. The store should have a supply of water to allow users to wash if exposed to any of the chemicals.

The person responsible for each store should be fully trained, as proper stock records are essential, with a regular careful check of stocks to ensure containers are not leaking and that the oldest stock is used first. If there is any pesticide in the store that has exceeded its use-by date, a sample should be sent to check its quality, as it may still be effective when used and avoid the cost of disposing of obsolete stock.

Waste disposal

Used containers and other materials exposed to pesticides must be carefully stored until they can be safely disposed of according to national procedures.

Conclusion

This chapter indicates that there are a number of diseases transmitted by insect vectors and that there are certain insecticides recommended for their control. The use of pesticides is regulated to reduce the possible adverse effects that can occur if the recommendations regarding the dosage and type of formulation applied are not adhered to.

References

Allsopp, R. (1990) A practical guide to aerial spraying for the control of tsetse flies (*Glossina* spp.). *Aerial Spraying Research and Development Project Final Report*, vol. 2. Natural Resources Institute, Chatham UK.

Aronson, N. E. (2007) Leishmaniasis in relation to service in Iraq/Afghanistan, US Armed Forces, 2001–2006. *Military Surveillance Monthly Report* **14**: 2–5.

Becker, N., Petric, D., Zgomba, M. and Boase, C., et al. (2010) *Mosquitoes and Their Control*. Springer, Dordrecht.

Bouwman, H., Reinecke, A. J., Coopman, R. M. and Becker, P. J. (1990) Factors affecting levels of DDT and metabolites in human breast milk from Kwazulu. *Journal of Toxicology and Environmental Health* **31**: 93–115.

Curtis, C. F., Malecela-Lazaro, M., Reuben, R. and Maxwell, C. A. (2002) Use of floating layers of polystyrene beads to control populations of the filaria vector *Culex quinquefasciatus*. *Annals of Tropical Medicine and Parasitology* **96, Suppl. 2**: S97–104.

Emerson, P. M., Lindsay, S. W., Walraven, G. E., et al. (1999) Effect of fly control on trachoma and diarrhoea. *Lancet* **24**: 1401–3.

Faiman, R., Cuno, R. and Warburg, A. (2009) Control of Phlebotomine sand flies with vertical fine-mesh nets. *Journal of Medical Entomology* **46**: 820–31.

Feachem, R. G. A., Phillips, A. A., Hwang, J., et al. (2010) Shrinking the Malaria map: progress and prospects. *The Lancet* **276**: 1566–78.

Gratz, N. G. (1999) Emerging and resurging vector-borne diseases. *Annual Reviews in Entomology* **44**: 51–75.

Groome, J. M., Martin, R. and Slatter, R. S. (1989) Advances in the control of public health insects by the application of water-based ultra low volume space sprays. *International Pest Control* **Nov/Dec**: 137–40.

Kgori, P. M., Modo, S. and Torr, S. J. (2006) The use of aerial spraying to eliminate tsetse from the Okavango Delta of Botswana. *Acta Tropica* **99**: 184–99.

Killick-Kenrick, R. (1999) The biology and control of phlebotomine sand flies. *Clinics in Dermatology* **17**: 279–89.

Lindh, J. M., Torr, S. J., Vale, G. A. and Lehane, M. J. (2009) Improving the cost-effectiveness of artificial visual baits for controlling the tsetse fly *Glossina fuscipes fuscipes*. *PLoS Neglected Tropical Diseases* 3(7).

Marquardt, W. (ed.) (2004) *Biology of Disease Vectors*. Academic Press, New York.

Matthews, G. A. (2000) *Pesticide Application Methods*, 3rd edition. Blackwell Publishing Ltd., Oxford.

Phanthumachinda, B. (1978) Problems of dengue haemorrhagic fever. Prevention and control in Thailand. *Asian Journal of Infectious Disease* **2**: 132–5.

Phuanukoonnon, S., Muellerz, I. and Bryan, J. H. (2005) Effectiveness of dengue control practices in household water containers in Northeast Thailand. *Tropical Medicine and International Health* **10**: 755–63.

Rayaisse, J. B., Tirados, I., Kaba D., et al. (2010) Prospects for the development of odour baits to control the tsetse flies *Glossina tachinoides* and *G. palpalis* s.l. *PLoS Neglected Tropical Diseases* 4(3).

Torr, S. J., Maudlin, I. and Vale, G. A. (2007) Less is more: restricted application of insecticide to cattle to improve the cost and efficacy of tsetse control. *Medical and Veterinary Entomology* **21**: 53–64.

WHO (1999) 'Principles for the assessment of risks to human health from exposure to chemicals.' Geneva, WHO (Environmental Health Criteria 210; available at: www.inchem.org/documents/ehc/ehc/ehc210.htm).

WHO (2001) 'Field testing and evaluation of insecticides for indoor residual spraying against Chagas vectors.' WHO/CDS/WHOPES/GCDPP/2001.1

WHO (2006) 'Pesticides and their application. For the control of vectors and pests of public health importance.' WHO/CDS/NTD/WHOPES/GCDPP/2006.1

WHO (2009) 'Dengue. Guidelines for diagnosis, treatment, prevention and control.' WHO/HTM/NTD/DEN/2009.1

WHO (2010a) 'Guidelines for laboratory and field testing of mosquito larvicides.' WHO/CDS/WHOPES/GCDPP/2005.13

WHO (2010b) 'Generic risk assessment model for insecticides used for larviciding.' WHO/HTM/NTD/WHOPES/2010.4 (web only) Further information on disease can be obtained from http://www.cdc.gov and http://www.afpmb.org. Information on the various insecticides mentioned in the text can be obtained from the manufacturer's webpage.

WHO (2011) Guidelines on Public Health Pesticide Management Policy in the WHO African Region. WHO, Geneva. WHO/HTM/NTD/WHOPES/2011.2

Chapter 2

Indoor Residual Spraying

The use of indoor residual spraying (IRS) really began following the development of the insecticide DDT during World War II and its use by the military to delouse the civilian population of Naples in early 1944. As a dust, it was applied at approximately 22 g per person. Over 3 million people were dusted and this kept the number of typhus cases down to just below 2,000 (Simmonds, 1959). The use of DDT in relation to tropical health was quickly recognised (Buxton, 1945). In Italy, following the success against typhus, a 5% DDT in kerosene mix was sprayed on walls in 5,896 buildings in the Tiber delta area of Isola Sacra. This was applied at 200 mg per square foot and was responsible for eliminating malaria in that area (Missiroli, 1946). Missiroli then developed a 5-year plan to eradicate malaria from the country by application of DDT residual spraying. As part of the planning, the island of Sardinia provided an experimental area with full-scale operations to control *Anopheles labranchiae*. The technical direction was by the International Health division of the Rockefeller Foundation from 1 October 1945 and the operations began in April 1946 by a special agency of the Italian High Commission for Hygiene and Public Health (Logan, 1953). A 5% DDT mix was sprayed at 2 g ai/m^2 of treated wall and ceiling surface and supplemented with a 2.5% DDT in fuel oil plus emulsifier application applied at 0.112 kg/hectare as a larvicide to watercourses. To carry out these operations, the island was divided into sectors of approximately 4.5 km^2. Difficulties in the terrain meant that over 250 vehicles were used together with animal transport, two helicopters, boats and rafts. However, by 1947, the Italian campaign was based entirely on residual spraying. At a cost of six billion lire (in 1949, US$ was equivalent of 625 lire) over 4.5 years the number of malaria cases fell from 78,173 in 1944 to 44 in 1950 and 9 in 1951 (Missiroli, 1950; Brown et al., 1976; Tognotti, 2009). The programme also effectively controlled flies and other domestic insects and lowered infant mortality rates and the incidence of intestinal infections (Barnard, 1949; Simmonds, 1959).

Integrated Vector Management: Controlling Vectors of Malaria and Other Insect Vector Borne Diseases, First Edition. Graham Matthews.
© 2011 John Wiley & Sons, Ltd. Published 2011 by John Wiley & Sons, Ltd.

Similar programmes were initiated in several countries and in 1955 the World Health Organisation launched a Global Malaria Eradication Programme, a campaign that set out to interrupt transmission in all endemic areas outside tropical Africa where the intensities of transmission were low to moderate (WHO, 1956). This campaign led to the successful reduction in malaria, with 37 of the 143 countries being free of malaria by 1978. Twenty-seven of these countries were in Europe and the Americas (Wernsdorfer, 1980). Elsewhere, the number of malaria cases declined in India from over 110 million in 1955 to less than 1 million in 1968, while the number of cases in Sri Lanka fell from 2.8 million in 1946 to only 18 reported in 1966.

In Africa, malaria eradication pilot projects were initiated in some countries, notably Benin, Burkina Faso, Burundi, Cameroon, Kenya, Liberia, Madagascar, Nigeria, Rwanda, Senegal, Uganda, and in the United Republic of Tanzania. Where these control programmes were used, there was a significant reduction of anopheline vector mosquitoes and malaria, but generally transmission of malaria was not interrupted (Kouznetsov, 1977) and the disease reappeared immediately control measures stopped, as in the Garki Project in Northern Nigeria (Molineaux and Gramicia, 1980). In that project, using propoxur and fenitrothion for IRS, biting rates per person would need to have been reduced to below 0.1 per person per night from as many as 42 bites per person (Jobin, 2010). As pointed out earlier by MacDonald (1957) and reiterated by Smith et al. (2007), success of malaria control depends partly on the basic reproductive number for malaria R_0. R_0 estimates vary considerably so where R_0 is highest, integration of different control techniques is crucial and is discussed in Chapter 6.

Nevertheless. IRS using DDT was scaled up in southern Africa and a few islands, such as Reunion, Cape Verde, Zanzibar and Sao Tome (Mabaso et al., 2004).

These programmes were not sustained due to lack of finance, concerns about insecticide resistance and community acceptance. Selection for resistance to DDT in mosquito populations was considered to be largely due to the insecticide getting into larval breeding areas, following its use in agriculture (Brown et al., 1976). However, Zahar (1984) suggested that in West Africa, apart from increased resistance to DDT, transmission of malaria was not interrupted following its use due to increased exophily of *An. gambiae* and *An. funestus*. Another problem encountered was the deposition of soot on the mud walls that reduced the effectiveness of DDT to no more than 2.5 months. Cooking with a fire is usually in a small separate house away from sleeping areas, so the presence of soot should not be a major problem.

There has been a resurgence of malaria since the 1970s (Figure 2.1). WHO abandoned eradication and relied on drug treatment and more recently the use of insecticide treated bed nets. However, it has been realised that IRS is an important means of reducing the vector and thus reduces the need for administering drugs to which the malaria parasite has developed resistance.

Since the 1970s, fear of harmful effects of persistent chemicals on the environment, such as DDT, due largely to its extensive use in agriculture, has

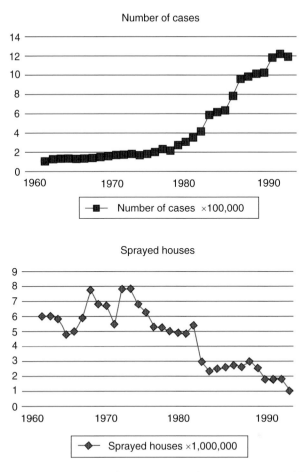

Figure 2.1 Impact of decrease in the number of houses sprayed with DDT on increase in malaria in South America (reproduced with permission from Roberts et al., 2010).

led to its inclusion in the Stockholm Convention that limits the use of persistent organic pollutants (POP) (Sadasivaiah et al., 2007). However, DDT remains on the list of insecticides recommended for IRS by WHO and its use has been accepted as a special case, exempting it from the Stockholm Convention for IRS. However, countries using DDT are required to have a licence and carry out a strict audit to ensure it is only used for IRS. Nevertheless, there has been much debate on whether it should be used. In supporting the use of DDT, Roberts et al. (2010) have pointed out that WHO and others relied too much on its toxic effect on mosquitoes and paid insufficient attention to how it worked, particularly the repellent effect of DDT deposits (Figures 2.2 and 2.3). Kennedy (1947) was among those who had pointed out that DDT operated also as an excitant and repellent, so that a high proportion of mosquitoes were repelled by a house sprayed with DDT and did not enter, or entered but departed without biting. Smith and Webley (1968) reported that 60-70% of

If house were unsprayed...

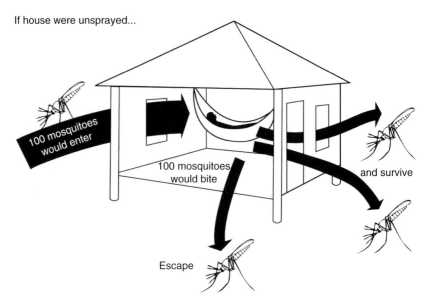

Figure 2.2   Depiction of mosquito behaviour in an unsprayed house for 100 mosquitoes (reproduced with permission from Roberts et al., 2010).

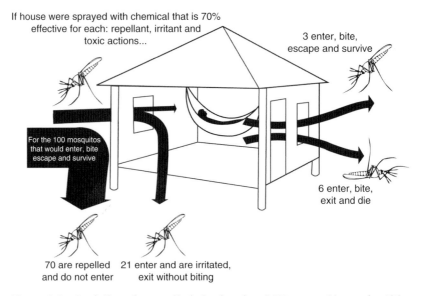

Figure 2.3   Depiction of mosquito behaviour in a DDT sprayed house for 100 mosquitoes (reproduced with permission Roberts et al., 2010).

An. gambiae and 70–80% of *Mansonia uniformis* were deterred from entering verandah-trap huts (Rapley, 1961; Smith, 1965) (Figure 2.4) for at least 4 months after the huts had been treated with DDT. Analysis of papers hung in the eaves of the experimental huts showed DDT moved as dust or vapour

(a)

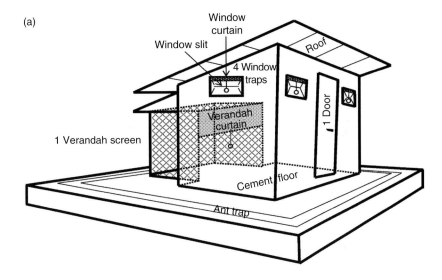

(b)

Figure 2.4 (a) Diagram showing an experimental hut; (b) photograph of an experimental hut; (c) cross-section of an experimental hut, for studies on mosquitoes.

from the treated wall surfaces. They considered that the deterrence was due to the outflow of DDT, which diminished from a maximum of 40 ng/cm²/day to a minimum of 0–2 ng/cm²/day, while a deposit of DDT built up on the untreated overhanging eaves until at 7 months after treatment it was 0.02–0.27 μg/cm², about ten times the concentration immediately after spraying.

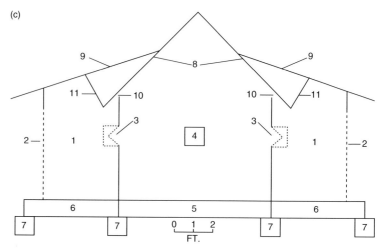

(c)

Diagram to show the basic design of the verandah-trap hut. 1, verandah trap; 2, removable screen of copper gauze; 3, window traps in walls to right and left; 4, window (closed) in wall facing the observer; 5, beam supporting wall of hut; 6, beam supporting verandah; 7, concrete pillar; 8, roof of hut; 9, roof of verandah; 10, eave; 11, partition of thatch cutting off upper part of verandah roof-space. It should be noted that not all the parts shown lie in the same vertical plane.

Figure 2.4 (*Continued*)

In South Africa, where DDT has been used most consistently for IRS, recent studies have evaluated how residents could be exposed to residues of DDT via different pathways (van Dyk et al., 2010). They examined air inside treated huts, dust, soil, food (leafy vegetables and chicken) and potable water 2 months after the huts were sprayed and compared the results with a similar area not sprayed with DDT situated outside the malaria control area. Detectable levels of DDT and its metabolites were found in most samples from the DDT treated houses, except chicken muscle samples, compared to the reference samples, indicating a combination of potential dietary and non-dietary pathways of uptake. Rogan and Chen (2005) reviewed published data on effects of DDT on humans from the 1960s and expressed concern that its use may cause preterm birth and early weaning, abrogating the benefit of reducing infant mortality from malaria. Bouwman et al. (2011) have pointed out that there are situations where DDT will provide the best achievable health benefit, but maintaining that DDT is safe ignores the cumulative indications of many studies.

In some countries, the strategy for IRS is to deploy it in endemic areas over three consecutive peak transmission seasons to reduce the burden of disease as the use of bed nets is scaled up, rather than adopting it as a key method supplemented with treated bed nets. The choice of area is dependent on surveillance data, so that IRS is focused on certain areas to contain major outbreaks of malaria and interrupt disease transmission.

In addition to controlling mosquitoes, in Latin America IRS has been used to control triatomine bugs, the vector of Chagas disease. It has also been used to control phlebotomine sand flies, where there have been outbreaks of leishmaniasis.

Apart from controlling target species, IRS usually reduces the presence of a number of other insect pests, including cockroaches, bed bugs and flies.

Equipment for indoor residual spraying

In the Global Programme, when WHO used IRS, it required all internal wall surfaces of dwellings to be sprayed with a dose of 2 g DDT ai/m^2 to control mosquitoes and reduce transmission of malaria. During the introduction of IRS, a number of problems were soon identified with the equipment. Compression sprayers were chosen as the sprayer does not require pumping while spraying, so that spray operators can give all their attention to correctly directing the nozzle at wall surfaces. However, existing compression sprayers used in agriculture invariably had a pump integral with a small opening of the tank and this had to be removed each time the sprayer was filled with insecticide spray. When the pump was re-inserted, soil or other material introduced into the tank caused nozzle blockages. There were other problems in the field, such as nozzle erosion, so WHO convened a meeting that subsequently led to the publication of specifications, which have been periodically revised (WHO, 1964, 1974, 1990, 2006).

The compression sprayer (Figure 2.5) that meets the specification has a pump with a T-shaped handle separate from the tank lid, a means of parking the lance when not in use, a foot rest and a hardened stainless steel 8002 even spray fan nozzle to minimise the effect of any abrasive particles in the spray liquid have on the nozzle. The lance has to have a trigger valve to start and stop spraying easily. This valve is required to have a lock-off position so that it is not possible to accidentally switch the spray on. A single shoulder strap of at least 50 mm wide is provided, so that the tank can easily be carried through doorways. The strap has to be made of material that does not absorb liquid, as in use it might get contaminated with insecticide. Testing the performance of the sprayers, in particular the durability of the tank subject to frequent pressurisation to 4 bar and depressurisation during field operations, was evaluated on a specially designed rig (Hall, 1955) in the USA and also in the UK (Figure 2.6). Similarly, the performance of the trigger valve without leakage was evaluated using initially a suspension of DDT wettable powder (WP), but later replaced by inert silica powder, used in the formulation of the WP.

A number of sprayer parts are subject to wear during use. These include various washers/gaskets to provide a liquid seal between components. These are often made of a synthetic elastomer, but are liable to swell and deform unless resistant to the insecticide used. Viton seals have been shown to be very effective. The pump plunger has been traditionally made of leather, as it can be moistened with oil if it dries out. Other designs of pumps may have an O-ring seal, but these are more prone to wear.

An 8002 nozzle provides a spray angle of 80 degrees and an output of 0.2 US gallons per minute at 40 psi pressure (757 ml/min at 3 bar), to treat adjacent swaths which overlap over the wall surface. However, the equivalent even

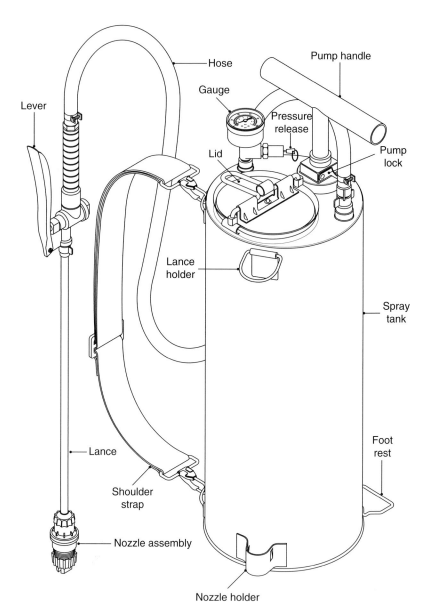

Figure 2.5 A compression sprayer to meet WHO specifications.

spray nozzle 8002E gives a more uniform distribution across the swath, but greater care is needed to avoid too great an overlap of adjacent swaths (Figures 2.7 and 2.8). Too great an overlap results in a strip receiving an overdose.

Hardened stainless steel (HSS) nozzle tips were chosen to mitigate the problem of erosion of the nozzle orifice. Traditionally, the HSS nozzle was 100% metal, but to conform to the ISO standard for a colour code on fan nozzles, the latest HSS nozzles have a metal tip embedded in plastic. Yellow

Figure 2.6 Compression sprayer under pressure test, simulating use of the equipment in the field for 12,000 pressurisations of the tank.

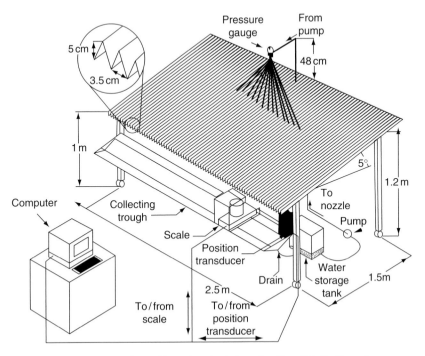

Figure 2.7 Example of a patternator to determine the distribution of spray across a swath.

indicates the flow rate for the 8002 nozzle that can also be referred to as FE80/0.8/3 (FE Flat fan; 80-degree spray angle; 0.8 litres per minute output at 3 bar pressure). It is now possible to purchase similar nozzles with a ceramic tip, embedded in a plastic nozzle body (Figure 2.9).

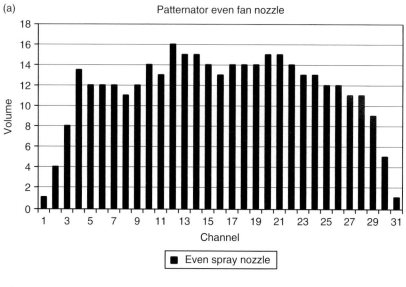

(a)   Patternator even fan nozzle

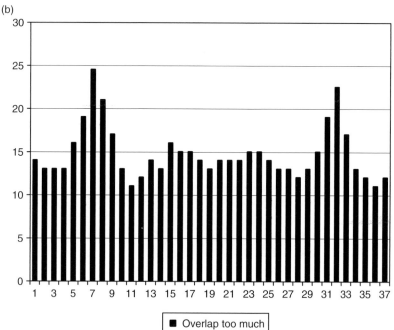

(b)

Figure 2.8 (a) Pattern of a single 8002E nozzle; (b) pattern when adjacent swaths overlap too much; (c) example of spray on a mud wall in Africa, showing overlap of adjacent swaths.

(c)

Figure 2.8 (*Continued*)

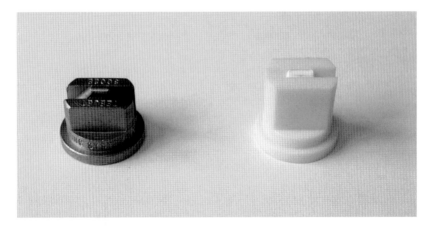

Figure 2.9   Hardened stainless steel and ceramic 8002E nozzle tips; the ceramic tip is embedded in a plastic body to protect the tip.

Deposits on wall surfaces were reported to vary (Gratz and Dawson, 1963; Smith and Webley, 1968) and efforts were made to overcome the variation in spray output due to the decrease in tank pressure while spraying (Hall and Taylor, 1962). Variability in deposits on walls has continued (Guillen et al., 1997), even with trained operators, which apart from variations in pressure at the nozzle during application, is also due to human error when some sections of individual walls are not treated correctly or missed.

A brass valve was used to control pressure at the nozzle but it could be adjusted by the user and was reported to be easily blocked due in part to its design, but also lack of efficient cleaning at the end of each spray session.

Tests were then carried out with a simple rubber disc that constricted the flow of spray at higher pressures (Lonergan and Hall, 1959), but it absorbed water. Discs soaked in a DDT suspension showed a substantial weight increase and did affect flow rate. A 9504E tip was considered more suitable with the disc flow regulator (Weathers et al., 1971). Field tests showed that although the disc flow regulator was simple to use, it clogged more frequently with debris and was easily lost when the equipment was cleaned (Fitzjohn and Stevens, 1963).

Some operators deliberately removed filters and any valves, as they considered these were likely to be the cause of blockages when WPs were applied. Blockages that did occur were either due to a poor quality formulation or inadequate cleaning of the sprayer at the end of the previous day. By not flushing clean water through the hose, lance and nozzle, powder dried overnight on the filter and may then have been the cause of a blockage. In practice, proper cleaning of the sprayer at the end of each day has often been ignored, so training programmes need to emphasise its importance.

Much later, a new design of control flow valve developed for knapsack sprayers was evaluated (Craig et al., 1993; Brown et al., 1997) and did maintain constant pressure and output at the nozzle, until the tank pressure was too low to open the valve. Further studies with a similar valve (Figures 2.10a, b) (Brown et al., 2003) confirmed the valve did provide a constant flow at the nozzle. Use of the valve (Figures 2.10c, d) was recommended by WHO in 2006. With the lower pressure at the nozzle (typically 1.5 bar), there is less bounce of droplets from wall surfaces, the average droplet size is larger and there are fewer droplets of less than 100 µm and thus less risk of the spray operator inhaling spray (Figure 2.11). Wolfe et al. (1959) showed that operator exposure was significantly reduced when spraying at 1.4 bar in comparison with 3.5 bar. and that it was less while spraying walls compared with ceilings.

In contrast to many agricultural sprayers, the lance is straight so that where the inside of the roof of some houses without a ceiling is high, an extra length can be added to extend the lance and enable spray to be applied close to the roof.

As the pressure gauge is one item that is frequently broken, the specification of the pump required it to increase the tank pressure by 1 lb per square inch (psi) with each full stoke of the pump, so that by counting the strokes, the operator could judge the pressure. Thus, to reach 4 bar pressure required about 58 pump strokes.

The current recommended equipment is specified in the *Guidelines* published by WHO (2006a). Those purchasing the equipment must ensure that they order sprayers that comply with the WHO specifications and are fitted with the correct nozzle and constant flow valve (CFV) and are supplied with sufficient spare parts.

Many houses do not have a ceiling, so the inside of the roof may be too high to spray with the standard lance provided with the sprayer. Such houses should be visited by two spray operators, one equipped with the standard lance, who is then responsible for spraying the walls, while the second operator has a sprayer on which the lance is fitted with an additional lance. The nozzle and CFV are unscrewed from the lance, the extra straight lance is screwed onto the

(a)

(b)

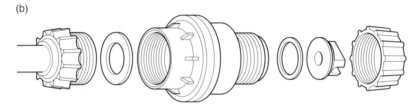

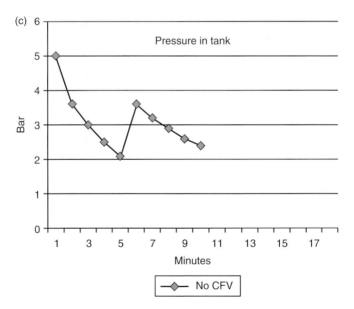

Figure 2.10 (a) Close-up of spray management valve (CFV); (b) exploded diagram of CFV; (c) pressure in tank of a compression sprayer without a CFV; (d) output of compression sprayer with three different pressures obtained using a 1.5 and 2 bar CFV. Note, spray stops when tank pressure is too low to open valve (Matthews, 2008).

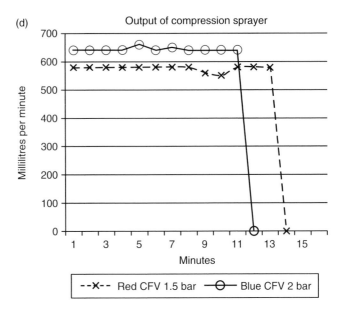

Figure 2.10 (*Continued*)

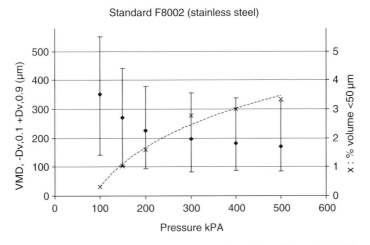

Figure 2.11 Change in droplet size and proportion of small droplets (<50 μm) at different tank pressures.

lance and the nozzle and CFV screwed back onto the end of the lengthened lance. This allows spray to be applied with the nozzle at a greater height, so that the inside of a roof can be treated. A different type of nozzle should not be fitted.

The compression sprayer has the advantage of requiring only manual effort to pressurise the sprayer and direct the spray, but the need to pump the tank pressure in the tank to 4 bar frequently, as the tank can be emptied in about 15 minutes, could possibly be obviated if a knapsack sprayer fitted with an electric pump and rechargeable batteries could be used. Electricity to recharge batteries is available in urban areas and many rural districts in contrast to the situation in the 1950s. However, this type of equipment has yet to be evaluated for vector control.

Spray volume

The volume to treat houses will depend on the type of wall surface. Mud walls are more absorbent than wood or cement surfaces. Traditionally, using the 8002 nozzle, a wall is sprayed with 40 ml/m^2 at 3 bar, but the volume, dose and droplet size change during the application as the tank pressure decreases. On non-porous walls, this volume of spray may flow down the wall surface, so a lower volume per square metre should be applied. Use of an 8001 nozzle has been suggested, but the smaller orifice is liable to be blocked more easily than with the 8002E nozzle, unless the water is very clean. The recommendation now is to use a CFV to ensure uniform flow at the nozzle and with a 1.5 bar CFV, the output is reduced and walls are sprayed at 30 ml/m^2. Applying this volume, the dose rate of product as active ingredient/m^2 remains the same and only the water volume is reduced by 25%. By inference, this means that standard sachets of insecticide designed to treat 250 m^2 of wall surface at 10 litres can then be diluted in 7.5 litres of water and thereby smaller tank sizes can be used. The smaller tank and less volume means that the operator is less tired carrying a lower weight.

Insecticides

WHO, through the pesticide evaluation programme (WHOPES), evaluates insecticides for IRS and has published guidelines for testing (WHO, 2006b). Trials were carried out in African huts situated at WHO centres, such as at Kaduna, Nigeria (Figure 2.12). WHO currently recommends 12 insecticides belonging to 4 chemical groups (1 organochlorine, 6 pyrethroids, 3 organophosphates and 2 carbamates). The recommendations are for products from specific manufacturers and are not valid unless linked to valid specifications. Having the correct formulation is extremely important, for example with WP, particle size and dispersing agents need to ensure the particles readily remain in suspension. The choice of insecticide in any one locality is determined from the recommended list (Table 2.1), by its safety to humans and the environment, and by the susceptibility of the vector species and its behaviour.

As the spray deposit on walls needs to be available to insects resting on the wall surface, the insecticide formulation needs to be a particulate such as a WP or micro-encapsulated formulation (CS). In micro-encapsulated formulations, the insecticide is enclosed in extremely small capsules and suspended in a liquid carrier, the aim being to increase the persistence of the insecticide by controlling its release from the spray deposit. The liquid formulations, such emulsifiable concentrates, tend to be absorbed into a wall and are less effective on the surface. Ideally where portable sprayers are

(a) (b) (c) (d)

Figure 2.12 WHOPES programme for testing new insecticides in Kaduna, Nigeria in the 1970s: (a) preparing spray; (b) proceeding to house; (c) leaving a treated house; (d) deposit on edge of doorway or window.

used, the insecticide formulation should be packaged in sachets so that the correct dose is added to individual sprayers.

The choice of insecticide will depend on results of bioassay tests to determine whether the local mosquito species are susceptible to the insecticide, both in the laboratory and on typical wall surfaces in the area being treated. The effectiveness of individual insecticides can vary between

Table 2.1 Insecticides currently recommended by WHO for Indoor Residual Spraying.

Insecticide	Type	Dose g/m^2
Bendiocarb	Carbamate	0.1–0.4
Propoxur	Carbamate	1–2
DDT	Organochlorine	1–2
Fenitrothion	Organophosphate	2
Malathion	Organophosphate	2
Pirimiphos methyl	Organophosphate	1–2
α-Cypermethrin	Pyrethroid	0.02–0.03
Bifenthrin	Pyrethroid	0.025–0.05
Cyfluthrin	Pyrethroid	0.02–0.05
Deltamethrin	Pyrethroid	0.02–0.025
Etofenprox	Pyrethroid	0.1–0.3
λ-Cyhalothrin	Pyrethroid	0.02–0.03

Note: A particulate formulation is required such as CS, WP.

species and can also be affected by the type of wall surface. Different types of mud wall, cement plaster or wooden surface will also affect the persistence of spray deposits.

It has been suggested that as only pyrethroids are used on bed nets, they should not be used for IRS (*Malaria World*, 20 May 2010), but whether use of the same insecticide on walls and bed nets increases the selection of resistance is not known, although it may well do so. Thus, ideally the insecticide used in IRS should not be the same as that used on bed nets, or at least the duration of its use should be limited, but at present the choice of insecticides with different modes of action suitable IRS is very limited. What is important is to have a sequence of different insecticides to minimise the risk of resistance to the insecticide being selected in the local populations of mosquitoes. In a different context, when red spider mites on cotton were detected to have resistance to dimethoate in Zimbabwe, it was decided to divide the country into three zones and acaricides with a particular mode of action were limited to one zone for 2 years and then moved to the next zone, reappearing in the first zone after a gap of 4 years (Duncombe, 1973). This programme was maintained very successfully for over 20 years.

A suggestion that different insecticides should be used, depending on the type of wall surface, is incorrect. The formulation and volume of spray are more critical in relation to the type of wall surface, so that a single chemical is used on an area-wide basis. This has not always been possible, with different donors supplying different insecticides within one country. At present, WHO is formulating a policy on insecticide resistance management.

Currently bendiocarb (Akogbeto et al., 2010) and pirimiphos methyl are the main alternatives to pyrethroid insecticides for IRS.

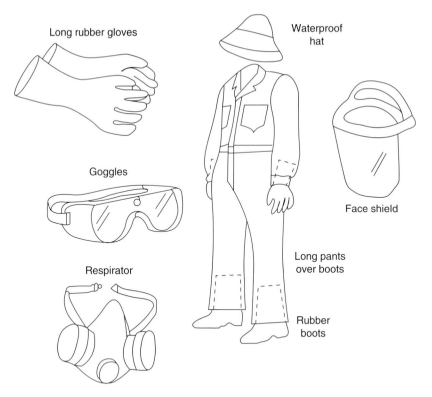

Figure 2.13 Items of protective clothing required when applying pesticides.

Operator exposure

The spray operator is the person who is nearest to the spray and must be protected while spraying (Figure 2.13). The operator must wear overalls, good non-absorbent footwear, gloves and a hat. The hat is to protect the head when spraying the surface of a ceiling. A cloth cover attached to the back of the hat can also protect the neck. The use of goggles or face mask has also been recommended and a mask to reduce the exposure of the mouth and nose to airborne droplets. The risk of inhalation of sprays is reduced by using a CFV set at a low pressure, typically 1.5 bar, in contrast to the pressure of 4 bar when the spray tank is fully pressurised and there is no CFV. Gloves should be washed before removal, so that any deposit on the glove is not transferred to the hands. Each product label should contain the minimum recommendations for protective equipment.

Resident exposure

The residents should not re-enter a house until the spray has had time to dry and the floor has been swept. They should be advised to avoid touching the wall surfaces and wash their hands if they do so before eating. Calculations have been made to assess the potential amount of insecticide that may be picked up by residents, to ensure that any insecticide approved for IRS does not exceed safety criteria. WHO has published a guide to making risk assessments for IRS (WHO, 2010).

Implementation of indoor residual spraying

Malaria vector control operations have to be targeted, treating only where and when necessary. The aim of IRS is to protect whole communities and is targeted at mosquito species that are endophilic, i.e. the mosquitoes that enter dwellings, so this requires virtually every house (>85%) within an area, to achieve the maximum impact on malaria transmission. The houses must be treated within a short period immediately before the transmission season. This may be annually in areas where there is a single rainfall period, but will be more frequent where transmission can occur throughout the year due to the rainfall pattern. Achieving this level of coverage and timing spraying correctly is crucial to real-ise the full potential of IRS, so it requires careful planning and sustained political commitment and financial resources over a prolonged period of several years, if not a decade. The implementation of IRS thus requires effective leadership and management, with operations directed by skilled professional staff, based on an analysis of local epidemiological data and a sound understanding of transmission patterns, vector behaviour and insecticide resistance status. The spray oper-ators, whether employed as members of a national team or locally recruited at regional or district/village level, need to receive practical training for which a certificate is awarded and valid for a defined period, such as three years. When planning and budgeting operations in the districts, attention needs to be given to the number, type and size of individual structures that need to be treated to derive the quantities of insecticide, equipment and storage facilities needed, as well as the number of qualifiedspray operators and supervisors needed to be trained and implement the programme. Training may also be needed to implement a geographical information system (GIS), which has been shown to facilitate malaria control in an area (Booman et al., 2000).

Village intervention teams

Traditionally, an IRS programme has been operated nationally by the Ministry of Health. Teams of spray operators are recruited to visit specific areas of a country and, with prior warning to the owners of houses to remove their furniture and any food, the team treats all the houses that are accessible. Where possible, all houses within a village should be treated, although access to some may not be possible. Many householders object to a total stranger entering their house, so an alternative approach is to train a small team to be responsible for the vector control operations within a village. Village (or Vector) Intervention teams (VIT) were used in village trials (Figures 2.14, 2.15

and 2.16) in Cameroon (Matthews et al., 2009) and in addition to the IRS treatments were also responsible for distribution and setting up the insecticide treated bed nets within a village and helping keep records of mosquitoes collected in exit traps (Figures 2.17a, c). Using VITs on a larger scale would require training a team of trainers who would be mobile and responsible for training the VITs in the villages, ensuring the appropriate supplies of inputs are distributed at the correct time to the District Health Centres and villages and subsequent monitoring within specific districts. Vector surveillance at village level is important data, which when collated at district and national

(a)

(b)

Figure 2.14 (a) A village house in Cameroon; (b) inside house, showing gaps in walls; (c) furniture removed from house in preparation for indoor residual spraying; (d) village water supply; (e) water for spraying in an old insecticide container (photo: Didier Baleguel). Note: these containers must never be used for storing water. In this case, the water was being used to apply a pesticide.

(c)

(d)

(e)

Figure 2.14 *(Continued)*

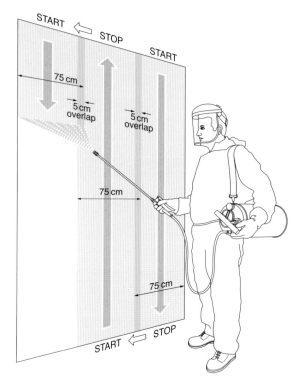

Figure 2.15 Diagram to show spraying a wall surface.

level, can provide a more detailed assessment of the effectiveness of a control programme and where improvements need to be made.

Involvement of local communities has been well established in the USA, where mosquito control districts supported through local taxation maintain a constant vigilance on mosquito populations within a county and take appropriate control actions as needed. There are over a 1,000 small municipal agencies, mostly in rural areas, and larger control units with annual budgets ranging from as little as $500 to as much as $24 million, with total annual expenditure by local government of about $200 million to control mosquitoes in the USA. Funding is derived from special county/municipal tax levies, as well as property assessments, and distributions of state taxes and federal grants (Conlon, 2011). With much higher standards of housing, control is largely with larviciding (Chapter 5), but space treatments (Chapter 3) are used when necessary. In some African countries, Village Health Teams (VHT) have been established to use rapid diagnostic tests supplied as a part of village health kits to ensure only those with malaria are treated with the artemisinin combination therapy (ACT). Previously, all cases of fever were often assumed to be due to malaria, but fever in young children may be due to other causes, hence the importance of accurate rapid screening to reduce the use of malarial drugs.

(a)

(b)

(c)

Figure 2.16 Diagram to show spraying a wall surface: (a) the sprayer;
(b) preparing the spray in the tank; (c) pumping the sprayer; (d) spraying a wall.

(d)

Figure 2.16   (*Continued*)

Planning programmes

Where IRS needs to carried out over a relatively short period prior to the rains, it is possible to do most of the planning and distribution of materials during the dry season. The planning team will need to liaise with the national authorities to ensure there is co-ordination of efforts between each district and that IRS fits in the national programme. In some areas, the programme is ideally organised on a regional basis, as in Southern Africa (Sharp et al., 2007).

Initially it will be important for the national organisation to define the districts where IRS will be done and appoint staff to be responsible for the districts and gain the support of the local population. In addition to the spray operators, it will be important to have a supervisor, who may travel to several villages during the spray programme, and other anciliary staff, such as store keepers.

In each district there will be a need to obtain maps of the appropriate localities within a district and collect data, as shown in Table 2.2. Information is also required on sources of local water and its availability for spraying and cleaning the equipment at the optimum time for treatment for each village.

Based on the information obtained, it will be important to calculate total insecticide required per village, and subsequently the total for all the villages within a district. At the same time it will be important to calculate the time needed to spray a house, and thus the number of sprayers needed to treat a village utilising a VIT or, if it is be treated by a national team of spray operators, how many will be needed for each village within the district. With the latter, time must be allocated for transport of the spray team from its base to the villages. Before any spraying can be done, the programme needs to be

(a)

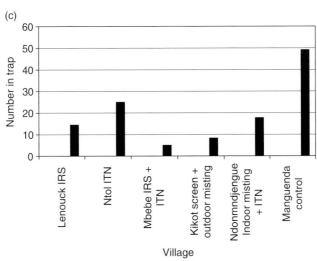

(b)

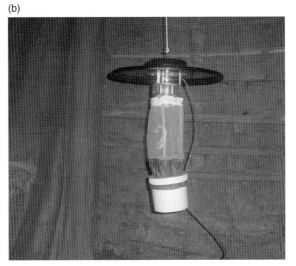

(c)

Figure 2.17 (a) Examining an exit trap; (b) light trap; (c) exit trap data from a village trial, Cameroon showing mean number of *Anopheles* mosquitoes per trap per night.

(d)

Figure 2.17 (*Continued*) (d) a simple resting box was placed outside the house (3 m) from the house wall. Resting boxes were placed outside the house at 6 pm. Mosquitoes resting in the box were easily collected using the mechanical hand aspirator at 6 am. There were three traps per house per night. In one trial, an average of 37 mosquitoes were sampled per box per day. The trap size was 45 cm long, 45 cm wide and 45 cm high. Cattle urine can be added to the box as an attractant (photo: Eliningaya John Kweka).

Table 2.2 Information that has to be collated for each district.

Location of villages	Co-ordinates determined using Global positioning systems (GPS) are important to define target dwellings and record information on a Geographical Information system (GIS) List names of villages and chief Number of dwellings per village (each dwelling should be assigned a number) – size, especially the wall area for each type of dwelling – dimensions of wall height and length, where possible
Population at risk of malaria in area of the district	Number of adults, pregnant women, children under 5, other children Population per village

discussed with the local authorities and each householder visited to ensure they understand what has to be done prior to the arrival of the spray team. In some areas it is useful to distribute a leaflet, explaining IRS and what the householder has to do. This should include explanation of the need to remove furniture, and other moveable items from the house during spraying, to allow easy access to wall surfaces. Any furniture that cannot be removed needs to be placed away from the wall and covered. Also any posters or other materials attached to walls should be removed. After spraying, the floor should be swept clean, and the walls left untouched as long as possible, with no resurfacing for at least six months. Inevitably there will be some posters or other pictures replaced on wall surfaces, but mosquitoes will still be affected by the deposits on the adjacent wall surface.

Insecticides

Insecticides currently recommended for IRS are shown in Table 2.1.

As the choice of insecticide has to be determined by the susceptibility of the local mosquito population, arrangements need to be made to collect sufficient mosquitoes to be used in bioassay tests. The manager of the overall programme can then make the appropriate arrangements to get the selected insecticide suitably packaged and transported to the district and villages before the dates set for spraying. This will probably necessitate ordering the insecticide several months before it is needed, so that any problems of delivery can be sorted out and delays in treatments avoided. The use of individual sachets for separate loads of spray is recommended, as it facilitates the mixing process and minimises the exposure of the operators to the insecticide. Where sachets are used, the entire contents must be emptied into the sprayer and empty sachets retained for return to the store for subsequent safe disposal.

Where other containers are used to prepare sprays, sachets should not be used, but it is important to have a suitable measure so that the correct amount is mixed and put into individual sprayers. Any insecticide on the measure must be rinsed with water that is added to the sprayer. When containers are empty, a small quantity of water should be added and the lid replaced. After shaking, the liquid is then poured into a separate large container, which is labelled 'Rinsate for subsequent use to dilute sprays'. This rinsing procedure needs to be repeated so that all used containers are washed three times – a procedure called 'triple rinsing'. This 'rinsate' and water used to wash out a sprayer at the end of each day can be stored overnight in a container and used to dilute sprays the following day. Alternatively, the washings can be used by spraying on the outside walls and into the eaves of houses. It is crucially important that any residual liquid containing insecticide is disposed of safely and does not contaminate any water supplies. Precautions are particularly important, when DDT is used, to ensure the insecticide is not used to spray crops.

Equipment required

In addition to determining the number of sprayers required in each district, it is important that sufficient spares and ancillary equipment, including containers for water and personal protective equipment (PPE) are also obtained. Having spares, including nozzles, washers and other items available, is important to ensure that routine maintenance can be done quickly.

Equipment is often taken in vehicles to different locations to treat houses. Care is needed to ensure that the individual sprayers and any spare parts are carefully packed so that they are not damaged during transit. When a number of sprayers are carried loose in the back of a truck, they can be easily damaged when travelling on rough roads by banging into each other. A simple frame should be used to hold each sprayer firmly during transit.

Storage

When the insecticide is taken to each district, it is essential that it is stored safely in a locked building, separate from any area with food and water supplies. There must be adequate ventilation and washing facilities for those who will responsible for its onward delivery to individual villages. Space is also needed to store the equipment and protect it from damage, such as chewing of hoses by rodents. A rat poison may be needed to avoid such damage. Stores will need to be inspected to determine their suitability and staff trained to maintain an inventory of the stocks. It is also important to ensure that only sufficient products are stored to avoid the problems associated with disposal of obsolete chemicals. Furthermore, precautions must be taken to ensure that insecticides intended for house spraying are not diverted illegally for agricultural use.

Training

Training of staff is a crucial part of any IRS programme and will require several types of training course. A course for management will involve those responsible for organising the activities in the different districts, so that actual spray operations take place at the correct time. They will need to be conversant with all the measurements that are needed during the planning of individual spray operations and subsequent monitoring. Where spray operations are devolved to individual districts and villages, the national authorities will require training programmes to 'Train the Trainers', who would then be responsible for training those who will do the actual spraying of houses. The trained 'trainers' can then visit each locality to train the individual spray operators and check that they can spray each house correctly. This will require practical training with sufficient time for individuals to be examined to ensure that they understand the whole process. This will include the wearing of the recommended Personal Protective Equipment (PPE), preparation and checking of the sprayer, the mixing of the insecticide, the spraying and subsequent refilling procedures, plus the procedure for washing the equipment at the end of each day. The operator should be able to change a nozzle if blocked or carry out simple repairs if necessary. More detailed maintenance can be carried out at a district workshop. WHO has published a manual on IRS, which should be translated into the local language to help operators understand the technique used and the procedures used to maintain the equipment. An animated version of the WHO *Manual on Indoor Residual Spraying* has been produced as a video on a DVD and is available for use on training courses.

Different training courses are required for the store keepers and those responsible for routine maintenance, to ensure that the supply and storage of the equipment is correctly carried out. At each store, one person should be responsible for maintaining the insecticides, equipment, spare parts and ancillary items safely and ensure that they are correctly audited. Donors should be requested to include a certain quantity of spare parts, determined by the number of sprayers being purchased. New equipment may fail if a small but important component breaks, so spares are essential. Facilities should also be provided so that equipment can be regularly checked and repaired when necessary. In many programmes, spares are not always available or not correctly fitted because inadequate practical training has been provided.

Monitoring

The monitoring of an IRS programme involves two parts. One is specific to the vector control aspect, while other monitoring of malaria incidence is part of a wider evaluation of the impact of interventions.

The IRS programme success is whether the population of mosquitoes is decreased and especially the number of mosquitoes that can infect people is sufficiently low to break the transmission of malaria.

Data needs to be collected on:

- The number of mosquitoes in an area. Ideally there will be some data from exit traps (Figure 2.17c) on at least one untreated house, but it will be important to have exit traps on a sample of treated houses and assess whether they have sporozoites, indicating whether they carried the malarial parasite. Odour-baited resting boxes have also been used to sample mosquitoes indoors and outside dwellings (Kweka and Mahande, 2009; Kweka et al., 2010) (Figure 2.17d). Tent traps have also been used (Govella et al., 2010).
- The programmes should be preceded by bioassays to assess the susceptibility of the local mosquitoes, and further bioassays conducted to check their susceptibility.
- Bioassays (Figure 2.18) will also be needed to check the persistence of the spray deposits in sample houses. Houses with different wall surfaces should be included if there is a range of building materials used in an area.

- In sample houses, it may be possible to arrange for mosquitoes to be collected at dawn from a sheet lid on the floor of a room, to determine how many entered the house and died before being able to leave it. By using a non-persistent pyrethrin spray, it is also possible to 'knock-down' any mosquitoes still alive and resting elsewhere in a room.
- In some trials, small light traps have been used, preferably provided with a source of carbon dioxide, such as dry ice, to improve the collection of

(a)

(b)

Figure 2.18 Bioassay on wall surfaces: (a) attaching a cone to a mud/wood wall; (b) putting mosquitoes in a cone fixed to a plank wall; (c) replacing the exposed mosquitoes into a clean container; (d) data from one trial.

(c)

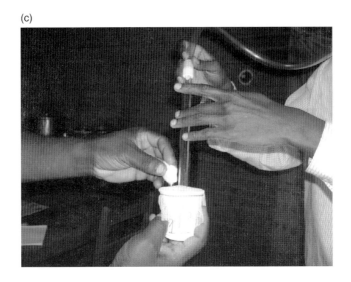

(d)

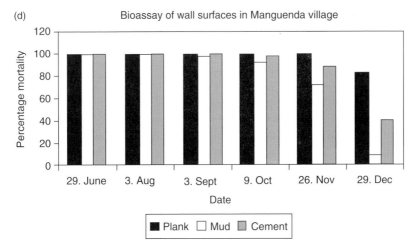

Figure 2.18   (*Continued*)

mosquitoes, but the results from such traps vary depending on the species. In Tanzania, Lines et al. (1991) used CDC traps alongside an occupied untreated bed net.
- Records at clinics and in villages of people reporting sickness will be needed in the wider context of reducing malaria in the overall programme, which will include insecticide treated bed nets and other interventions. Deployment of a rapid diagnostic method to check whether a sickness/ fever is due to malaria or other illness will enable the impact on incidence of malaria to be more clearly defined.

Environmental assessment

Donor agencies now require environmental assessments, due to the concern that use of pesticides may have an adverse impact not only on human health but also on other organisms in the environment. IRS is essentially confined to inside dwellings, although in some places spray operators may treat outside walls around doorways, windows and the eaves of houses, especially if access is denied. After spraying, householders are instructed to sweep the floor so floor dust contaminated with spray will be brushed out of a doorway. Animals, especially chickens, may be exposed to some of this dust. This has generally not been considered a problem, except where DDT has been used, because of its persistence in the environment.

Studies in South Africa have shown that although the inside of houses is sprayed, DDT may reach the outdoor environment from possible spillages during application and via airborne droplets and dust, especially as householders sweep the floor. Residues of DDT and its metabolites have been detected in wild fish, birds and especially in domestic chickens, which may be an important route of uptake by those living in sprayed houses (Barnhoorn, 2009), but generally the amounts detected are extremely low. In a study in Brazil (Vieira et al., 2001), chicken eggs had on average of twice the maximum residue level (MRL) (FAO), but considering just egg consumption, DDT intake was estimated to be 13 times less than the US EPA maximum reference dose (5×10^{-4} mg/kg bw/day), suggesting that this may be of concern only where eggs are a significant part of the diet. However, although DDT is stored in body fat, it is metabolised and excreted from the human body once it reaches a certain level.

Evaluation

As noted earlier in this chapter, historically there have been many examples where IRS has had a highly significant impact on populations of *Anopheles* mosquitoes and reduced the incidence of malaria. Careful records need to be kept of all future IRS operations, specifically to know exactly how many houses are treated, when and with which insecticide, together with monitoring data so that if there is any decline in impact, the reasons can be determined and arrangements made to correct the problem.

Undoubtedly IRS programmes have been highly successful in malaria reduction throughout the world, but following the decision by WHO to stop spraying houses with DDT due to resistance in the 1970s, there have been very few well-run trials that were considered suitable to include in a Cochrane study of the impact of IRS on malaria and quantify the effects of IRS in areas with different malaria transmission.

From the six studies included by Pluess et al. (2010), IRS reduced malaria transmission in young children by half compared to no IRS in Tanzania in an area where people are regularly exposed to malaria. IRS also protected all age groups in India and Pakistan, where malaria transmission is more unstable and where more than one type of malaria is found. No firm conclusions could be drawn in their review, when IRS was compared with insecticide treated nets (ITN).

Economics

The cost of IRS is often considered to be too great, due to the logistics of national campaigns, when large numbers of temporary staff require training and have to be transported around the country to do the actual spray treatments. However, costs vary between countries and improved formulation of insecticides is increasing their persistence on wall surfaces.

In one study in southern Africa, two separate programmes in rural and peri-urban areas were compared and showed that while the economic cost per person per year using a carbamate insecticide (bendiocarb or propoxur, both much more expensive than DDT or a pyrethroid) was $3.48 in rural and $2.16 in the peri-urban area. Adding 39 and 31%, respectively, to cover project management, monitoring and surveillance, the financial costs were $3.86 and $2.41, respectively, thus showing that an IRS programme can reduce suffering due to malaria, provided there are financial support, political will, collaborative management and training and community involvement in place (Conteh et al., 2004). Comparison of IRS costs with the use of bed nets is not easy, as there have been different ways in distributing the nets, including distribution of free nets. Yukich et al. (2008) suggested a variation between $1.21 in Eritrea to $6.05 in Senegal. De Allegri et al. (2010) clearly show that in addition to $5.78 for the net, distribution costs added $2.29 per net, so overall cost per net was $8.08. The cost of bed nets per year then depends on the life of the net and whether it has to be retreated.

Unfortunately, in Africa, as pointed out by Guyatt et al. (2002), the cost of protecting a household with ITNs would be equivalent to sending three children to primary school for a year. Thus the aspiration by poor rural homesteads to protect themselves with ITNs is not compatible with their ability to pay. Hence the importance of the Global Fund, World Bank, United States Agency for International Development (USAID) and other donors in providing funds to combat malaria.

In Ghana, the AngloGold Ashanti mining company's Obuasi malaria control programme calculated that it needed 116 spray operators to cover the intervention site in 5 months. All the recruits were examined to check that they met the physical requirements for the spraying tasks, followed by interviews for literacy and numeracy, so that those trained were fit, healthy, articulate and able to complete the required paperwork. The spraying programme required 18,560 kg of insecticide per spray round, 60 sprayers, 8 vehicles and 9 trailers with associated equipment. Between January 2005 and September 2007, the number of reported malaria cases declined by 73%, with an average reduction of 4,550 cases per month, costing $1.7 million for the first year, with an annual budget of $1.3m thereafter.

Box 2.1   Key Stages in IRS Programme

Identification of area requiring mosquito control
Mapping and survey to determine population at risk of malaria,
number of houses and area of wall surfaces to be treated
Bioassays - determination susceptibility of mosquitoes
Decision of insecticides and equipment/costs
Purchasing
Notification to householders
Training of all involved
Distribution of insecticide and equipment to districts
Implementation of IRS. Subsequent monitoring and evaluation

To reduce the cost of IRS applications, a trial in Kenya suggested that treatment of 25% of the houses still had an impact on malaria transmission. However, to have a major impact reduction on malaria, it has always been argued that at least 80% of houses should be treated (Zhou, 2010).

It has been suggested that IRS is best suited to areas where there is a limited period during the year when people are exposed to malaria transmission. In such areas, people may be less inclined to sleep under a net, and authorities may be disinclined to distribute nets to an area where they are only needed for a short period if there is an epidemic of malaria. It has also been suggested that the cost of IRS can be reduced if an integrated Malaria Early Warning and Response System (MEWS) is introduced to provide timely and accurate information on the severity of the mosquito population. Thus, the average marginal cost per case prevented was never more than $5.19, with 100% IRS coverage in medium, high or epidemic transmission years but was over $20-$500 per case in low transmission years (Worrall et al., 2008).

The cost of interventions is discussed further in Chapter 6. This chapter has set out information to enable those responsible for an IRS to obtain the basic information they need. Key steps are shown in Box 2.1.

Conclusion

IRS was a key tool against the Anopheline mosquitoes in the early days of DDT use. The enormous benefits of reducing malaria by using IRS have been largely forgotten, as DDT was banned for environmental reasons when it was overused in agriculture and resistance was perceived to be a major disadvantage to its use. Nevertheless, IRS remains a part of our toolkit in the fight against malaria, but for benefits to be achieved from using the technique, it will require greater commitment by governments and private sector participation to establish the infrastructure and training of staff to ensure spray programmes are correctly managed and sustained.

References

Akogbéto, M. C., Padonou, G. G., Gbénou, D., Irish, S. and Yadouleton, A. (2010) Bendiocarb, a potential alternative against pyrethroid resistant *Anopheles gambiae* in Benin. *West Africa Malaria Journal* **9**: 204.

http://www.anglogold.com/subwebs/InformationForInvestors/Reports07/ReportToSociety07/malaria-obuasi.htm

Barnard, C. I. (1949) *The Rockefeller Foundation – A Review for 1949* (Malaria in Sardinia). The Rockefeller Foundation, New York.

Barnhoorn, I. E. J., Bornman, M. S., van Rensburg, C. J. and Bouwman, H. (2009) DDT residues in water, sediment, domestic and indigenous biota from a currently DDT-sprayed area. *Chemosphere* **77**: 1236–41.

Booman, M., Durrheim, D. N., La Grange, K., et al. (2000) Using a geographical information system to plan a malaria control programme in South Africa. *Bulletin of the World Health Organization* **78**: 1438–44.

Bouwman, H., van den Berg, H. and Kylin, H. (2011) DDT and malaria prevention: addressing the paradox. *Environmental Health Perspectives* doi: 10.1289/ehp.1002127 (available at http://dx.doi.org)

Brown, A. W. A., Haworth, J. and Zahar, A. R. (1976) Malaria eradication and control from a global standpoint. *Journal of Medical Entomology* **13**: 1–25.

Brown, J. R., Zyzak, M. D., Callahan, J. H. and Thomas, G. (1997) A spray management valve for hand-compression sprayers. *Journal of the American Mosquito Control Association* **13**: 84–6.

Brown, J. R., McAuliffe, D. D., Smith, K. T., Beavers, G. M. and Presley, S. M. (2003) A constant flow valve for hand-compression hydraulic sprayers. *Journal of the American Mosquito Control Association* **19**: 91–3.

Buxton, P. A. (1945) The use of the new insecticide DDT in relation to the problems of tropical medicine. *Transactions of the Royal Society of Tropical Medicine and Hygiene* **38**: 367–93.

Conlon, J. (2011) Control of mosquitoes in the United States. *Outlooks on Pest Management* **22**: 32–5.

Conteh, L., Sharp, B. L., Streat, E., Barreto, A. and Konar, S. (2004) The cost and cost-effectiveness of malaria vector control by residual insecticide house-spraying in southern Mozambique: a rural and urban analysis. *Tropical Medicine and International Health* **9**: 125–32.

Craig, I. P., Matthews, G. A. and Thornhill, E. W. (1993) The spray management valve: a method to reduce spray drift and improve application accuracy. *ANPP/BCPC 2nd International Symposium on Pesticides Application* 341–8.

De Allegri, M., Marschall, P., Flessa, S., et al. (2010) Comparative cost analysis of insecticide-treated net delivery strategies: sales supported by social marketing and free distribution through antenatal care. *Health Policy and Planning* **25**: 28–38.

Duncombe, W. C. (1973) The acaricide spray rotation for cotton. *Rhodesian Agricultural Journal* **70**: 115–8.

Fitzjohn, R. A. and Stevens, P. A. (1963) Field test of rubber disc flow-regulators on compression sprayers. *Bulletin of the World Health Organisation* **29**: 375–86.

Govella, N. J., Moore, J. D. and. Killeen, G. F. (2010) An exposure-free tool for monitoring adult malaria mosquito populations. *American Journal of Tropical Medicine and Hygiene* **83**: 596–600.

Gratz, N. G. and Dawson, J. A. (1963) The area distribution of an insecticide (fenthion) sprayed inside the huts of an African village. *Bulletin of the World Health Organisation* **29**: 185–96.

Guillen, G., Diaz, R., Jemio, A., Cassab, J. A., Pinto, C. T. and Schofield, C. J. (1997) Chagas disease vector control in Tupiza, southern Bolivia. *Memorias do Instituto Oswaldo Cruz* **92**: 1–8.

Guyatt, H. L., Ochola, S. A. and Snow, R. W. (2002) Too poor to pay: charging for insecticide-treated bed nets in highland Kenya. *Tropical Medicine and International Health* **7**: 846–50.

Hall, L. R. (1955) Suggested techniques, equipment and standards for the testing of hand insecticide-spraying equipment. *Bulletin of the World Health Organisation* **12**: 371–400.

Hall, L. B. and Taylor, J. E. (1962) Regulated flow of insecticides. *Bulletin World Health Organisation* **27**: 279–81.

Jobin, W. (2010) *A Realistic Strategy for Fighting Malaria in Africa*. Blue Nile Monograph. Boston Harbor Publishers, Colorado.

Kennedy, J. (1947) The excitant and repellent effects of mosquitoes of sub-lethal contacts with DDT. *Bulletin of Entomological Research* **37**: 593–607.

Kouznetsov, R. l. (1977) Malaria control by application of indoor spraying of residual insecticides in tropical Africa and its impact on population health. *Tropical Doctor* **7**: 81–91.

Kweka, E. J. and Mahande, A. M. (2009) Comparative evaluation of four mosquitoes sampling methods in rice irrigation schemes of lower Moshi, northern Tanzania. *Malaria Journal* **8**: 149.

Kweka, E. J., Mwang'onde, B. J. and Mahande, A. M. (2010) Optimization of odour-baited resting boxes for sampling malaria vector, *Anopheles arabiensis* Patton, in arid and highland areas of Africa. *Parasites and Vectors* **3**: 75.

Lines, J. D., Curtis, C. F., Wilkes, T. J. and Njunka, K. J. (1991) Monitoring human-biting mosquitoes (Diptera, Culicidae) in Tanzania with light traps hung beside mosquito nets. *Bulletin of Entomological Research* **81**: 77–84.

Logan, J. A. (1953) *The Sardinian Project*. John Hopkins Press, Baltimore; Oxford University Press, London. 415 pp.

Lonergan, R. P. and Hall, L. B. (1959) The regulation of flow through residual spray nozzles. *Bulletin World Health Organisation* **20**: 961–71.

Mabaso, M. L. H., Sharp, B. and Lengeler, C. (2004) Historical review of malaria control in southern Africa with emphasis on the use of indoor residual spraying. *Tropical Medicine and International Health* **9**: 846–56.

Macdonald, G. (1957) *The Epidemiology and Control of Malaria*. Oxford University Press, Oxford. 201 pp.

Matthews, G. A., Thornhill, E. W. and Dobson, H. (2008) The compression sprayer and its use in vector control. *Aspects of Applied Biology* **84**: 337–42.

Matthews, G. A., Dobson, H. M., Nkot, P. B., Wiles, T. L. and Birchmore, M. (2009) Preliminary examination of integrated vector management in a tropical rain forest area of Cameroon. *Transactions of the Royal Society of Tropical Medicine and Hygiene* **103**: 1098-104.

Missiroli, A. (1946) La malaria nel 1945 e previsioni per il 1946. *Conference at Instituto Superiordi SanitC*, February 1946. Rome.

Missiroli, A. (1950) The control of domestic insects in Italy. *American Journal of Tropical Medicine* **30**: 773-83.

Molineaux, L. and Gramicia, G. (1980) *The Garki Project*. WHO, Geneva.

Pluess, B., Tanser, F. C., Lengeler, C. and Sharp, B. L. (2010) Indoor residual spraying for preventing malaria. *Cochrane Database of Systematic Reviews*, Issue 4. Art. No.:CD006657.DOI:10.1002/14651858.CD006657.pub2.

Rapley, R. E. (1961) Notes on the construction of experimental huts. *Bulletin of the World Health Organisation* **24**: 659-63.

Roberts, D., Tren, R., Bate, R. and Zambone, J. (2010*) The Excellent Powder: DDT's Political and Scientific History*. Dog Ear Publishing, Indianapolis.

Rogan, W. J. and Chen, A. (2005) Health risks and benefits of bis(4-chlorophenyl)-1,1,1-trichloroethane (DDT). *Lancet* **366**: 763-73.

Sadasivaiah, S., Tozan, Y. and Breman, J. G. (2007) Dichlorodiphenyltrichloroethane (DDT) for indoor residual spraying in Africa: how can it be used for malaria control. *American Journal of Tropical Medicine and Hygiene* **77 (Suppl. 6)**: 249-63.

Sharp, B. I., Kleinschmidt, I., Streat, E., et al. (2007) Seven years of regional malaria control collaboration – Mozambique, South Africa, and Swaziland. *American Journal of Tropical Medicine and Hygiene* **76**: 42-7.

Simmonds, S. W. (1959) The use of DDT insecticides in human medicine. In Simmonds, S. W. (ed.), *DDT, the Insecticide dichlorodiphenyltrichloroethane and its significance*. Vol II *Human and Veterinary Medicine*. Birkhauser Verlag, Basel and Stuttgart.

Smith, A. (l965) A verandah-trap hut for studying the house-frequenting habits of mosquitoes and for assessing insecticides. I. A description of the verandah-trap hut and of studies on the egress of *Anopheles gambiae* Giles and *Mansonia uniformis* (Theo.) from an untreated hut. *Bulletin of Entomological Research* **56**: 161-7.

Smith, A. and Webley, D. J. (1968) A verandah-trap hut for studying the house-frequenting habits of mosquitoes and for assessing insecticides: The effect of DDT on behaviour and mortality. *Bulletin of Entomological Research* **59**: 33-46.

Smith, D. L., McKenzie, F. E., Snow, R. W. and Hay, S. I. (2007) Revisiting the basic reproductive number for malaria and its implications for malaria control. *PLoS Biology* **5(3)**: e42.

Tognotti, E. (2009) Program to eradicate malaria from Sardinia 1946-50. *Emerging Infectious Diseases* **15**: 1460-6.

Van Dyk, J. C., Bouwman, H., Barnhoorn, I. E. J. and Bornman, M. S. (2010) DDT contamination from indoor residual spraying for malaria control. *Science of the Total Environment* **408**: 2745-52.

Vieira, E. D. R., Torres, J. P. M. and Malm, O. (2001) DDT environmental persistence from its use in a vector control program: a case study. *Environmental Research* **86**: 174-82.

Weathers, D. B., Taylor, J. W. and Jensen, J. A. (1971) A re-examination of the Disk Flow Regulator. *Bulletin of the World Health Organisation* **44**: 847-54.

Wernsdorfer, W. H. (1980) The importance of malaria in the world. In: Kreier, J. P., ed., *Malaria*, vol. 1. Academic Press, New York. pp. 1-93.

WHO (1956) WHO 'Expert Committee on Malaria', sixth report. Technical Report Series 123. WHO, Geneva. http://whqlibdoc. who.int/malaria/WHO_Mal_180.pdf.

WHO (1964) 'Equipment for vector control.' WHO, Geneva.

WHO (1974) 'Equipment for vector control', 2nd edn. WHO, Geneva.

WHO (1990) 'Equipment for vector control', 3rd edn. WHO, Geneva.

WHO (2006) 'Equipment for vector control specification guidelines.' WHO, Geneva.

WHO (2006) 'Guidelines for testing mosquito adulticides for indoor residual spraying and treatment of mosquito nets.' WHO/CDS/NTD/WHOPES/GCDPP/2006.3.

WHO (2010) 'Generic Risk Assessment Model for Indoor Residual Spraying of Insecticides.' WHO/HTM/NTD/WHOPES/2010.5 (web only).

Wolfe, H. R., Walker, K. C., Elliott, J. W. and Durham, W. F. (1959) Evaluation of the health hazards involved in house-spraying with DDT. *Bulletin of the World Health Organisation* **20**: 1-14.

Worrall, E., Connor, S. J. and Thomson, M. C. (2008) Improving the cost-effectiveness of IRS with climate informed health surveillance systems. *Malaria Journal* **7**: 263.

Yukich, J. O., Lengeler, C., Tediosi, F., et al. (2008) Costs and consequences of large-scale vector control for malaria. *Malaria Journal* **7**: 258.

Zahar, A. R. (1984) Vector control operations in the African context. *Bulletin of the World Health Organisation* **62 (Suppl)**: 89-100.

Zhou, G., Githeko, A. K., Minakawa, N. and Yan, G. (2010) Community-wide benefits of targeted indoor residual spray for malaria control in the Western Kenya Highland. *Malaria Journal* **9**: 67.

Chapter 3

Space Treatment

In contrast to applying a residual deposit on wall surfaces, a space treatment aims to create a cloud of extremely small droplets that remain airborne and are collected by flying insects. The use of a space treatment against mosquitoes dates back to the use of the simple manually pumped 'Flit Gun' to atomise a solution of natural pyrethrins in a light oil (Figure 3.1a). The insecticide solution in a small tank is sucked up by each suction stroke of the pump and then forced out of a small aperture, combined with air, during the compression stroke of the pump. This action creates the fine spray mist or aerosol. Brown et al. (1998) gave data on six versions of the Flit Gun. The Flit Gun was used in Africa to treat vehicles that had to enter a roadside building when leaving a tsetse infected area, to ensure that the tsetse fly was not taken into a new area.

The use of this low cost applicator declined when the ubiquitous 'Aerosol' can was developed with the insecticide pressurised inside the can, but is nevertheless a useful low cost applicator. The pressure pack (Figures 3.1b, c) cut out the variation in output and droplet size caused by the variation in use of the air pump. However, pressure packs are expensive and their use is limited to treating inside rooms of more affluent houses. Special aerosols are also used regularly on aircraft to minimise the risk of transporting malaria vectors and other insects (Figure 3.1d). Other methods of space treatment evolved from equipment used for spraying paint equipment and the use of fogs to camouflage ships during World War II.

Although not technically a 'space spray', in many parts of the tropics the mosquito coil has provided an inexpensive way of repelling mosquitoes from around the sleeping area. It consists of an extruded ribbon of wood dust, starch and colouring agent, together with an insecticide, traditionally natural pyrethrin. Each coil is usually 12 g in weight and should burn continuously for 7.5 hours (Figure 3.2). Chadwick (1975) suggested that the effect of the smoke from a mosquito coil was initially deterrence, and then to expel the mosquito from the area, thus reducing the biting of those sleeping in the area.

Integrated Vector Management: Controlling Vectors of Malaria and Other Insect Vector Borne Diseases, First Edition. Graham Matthews.
© 2011 John Wiley & Sons, Ltd. Published 2011 by John Wiley & Sons, Ltd.

Large-scale space treatment against mosquitoes can be divided into three groups:

1) The use of manually carried equipment that can be taken inside dwellings or areas inaccessible to larger equipment;
2) the use of vehicle mounted equipment for area-wide treatment, usually in urban areas; and
3) aerial applications when very large areas (thousands to tens-of-thousand of hectares) need to be treated to reduce a major threat from mosquitoes. (Except for the large, locally funded programmes in the United States, this method is usually only conducted in response to natural disasters and disease epidemics.)

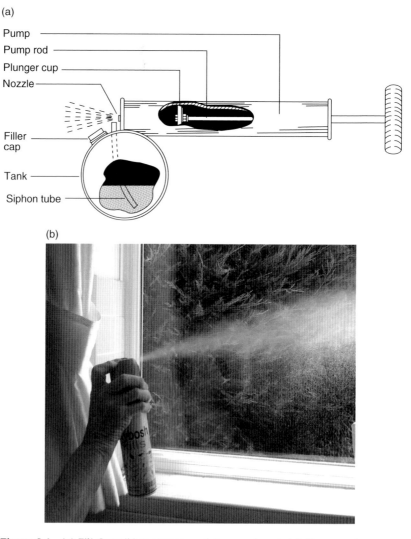

Figure 3.1 (a) Flit Gun; (b) pressure pack (aerosol can); (c) diagram of pressure pack; (d) disinfection of aircraft cabin to prevent mosquitoes travelling to a different country (photo: Roy Bateman).

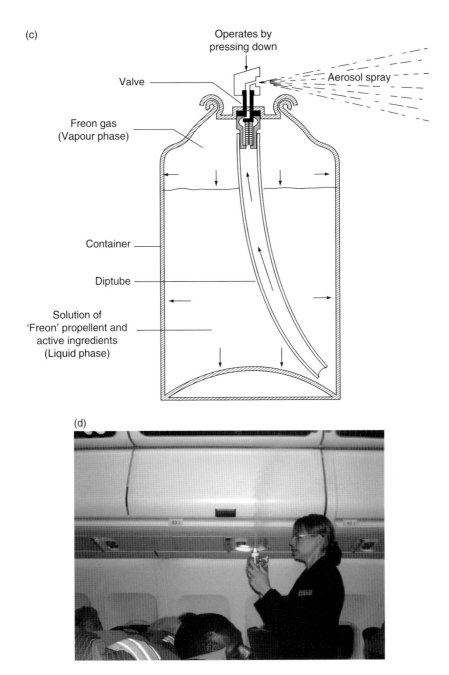

(c)

Operates by
pressing down

Valve

Aerosol spray

Freon gas
(Vapour phase)

Container

Diptube

Solution of
'Freon' propellent and
active ingredients
(Liquid phase)

(d)

Figure 3.1 (*Continued*)

Vehicle mounted equipment is often used in urban areas, especially in Southeast Asia, where it would be too costly and extremely difficult to have access to a very large number of residences to attempt to use an indoor residual spray (Gratz, 1999). Vehicle mounted fogging equipment is also used widely in the USA, where mosquito control programmes more often target

Figure 3.2 Mosquito coil used to irritate mosquitoes and repel them in Vietnam (photo: Roy Bateman).

high populations of nuisance mosquitoes rather than solely focusing on disease vectors. However, recent outbreaks of West Nile, Dengue, St. Louis Encephalitis and Eastern Equine Encephalitis viruses highlight the importance of maintaining active mosquito control capabilities. In areas with an epidemic of dengue fever, authorities may require an outdoor space treatment to reduce the principal vector *Aedes aegypti* rapidly and thus minimise the spread of the disease (Pant et al., 1971). Aerial applications to control mosquitoes in the USA are commonly conducted in emergency situations after devastation following hurricanes, with swaths typically up to 300–500 m wide (Mount et al., 1996) that have occurred at various parts of the east and southern coasts. Following some deaths due to West Nile fever in the late 1990s, areas of New York City were sprayed aerially with malathion and resmethrin. In June 2010, helicopters were still being used to apply larvicide over marshy areas in the Bronx, Staten Island and Queens areas.

In Europe, the aerial application of pesticides is widely discouraged and only permitted under specific circumstances, usually when no effective alternatives are available. Pesticide registration is governed by two pieces of legislation, one for plant protection products (covering aerial crop spraying), Directive 91/414/EEC (to be replaced by Regulation 1107/2009 from June 2011), and the other for biocidal products (including vector control), Directive 98/8/EC. One essential element of the product dossier is the environmental risk assessment, which must demonstrate that the use of the product does not cause any unacceptable risks to non-target animals (such as beneficial insects, aquatic invertebrates), humans (e.g. bystander exposure via drift) or the environment. Once registered, the use of pesticide products is addressed separately by the *Framework Directive for the Sustainable Use of Pesticides*, 2009/128. During the discussions leading to the final version of the text of this Directive, amendments were introduced that included a complete prohibition of aerial spraying in Europe, and which broadened the scope of the legislation to cover both plant protection and biocidal products. In the end, the scope was restricted to plant protection and aerial spraying remains permissible, but only with local derogations where the Member State concerned must inform the European Commission of their reasons to allow aerial spraying.

For vector control, many areas that are infested with mosquitoes, such as marsh land in river valleys or the rice fields of northern Italy and Greece, are not readily accessible for ground-based application (e.g. from vehicle mounted foggers), and aerial application remains the only viable means of effective treatment. Even so, specific legislation sometimes exists, even at the level of the city, or county council, for example, the individual Lande, or States, in Germany, which governs which products may be applied, by whom, and when. There is an increasing awareness that certain mosquito species are (re-) gaining a foothold in Europe, notably the Tiger mosquito, *Ae. albopictus*. This species was responsible for the recent outbreak of Chikungunya in the Ravenna area of northern Italy (almost 250 human cases reported in 2007 (Weissmann, 2008)), and recent monitoring detected the species in The Netherlands. Europe has also seen recent outbreaks of West Nile Virus, most commonly associated with *Culex* species (although the Tiger mosquito may also carry the virus), one in northern Italy (9 human cases reported in 2008, Barzon et al., 2009) and others in Romania (2 deaths reported in Bucharest in August 2010, Anon, 2010a) and Macedonia, northern Greece, where 134 human cases and 4 deaths were already known at the time of writing (August 2010, Anon 2010b). The outbreak in Greece has resulted in emergency provisional 120-day approval for aerial mosquito adulticiding (larviciding by helicopter was already approved).

Most aerial application in Europe is associated with application of larvicides, for which helicopters have been used (e.g. over the rice fields in Italy and Greece and in some areas along the Rhine Valley in Germany) (Chapter 5).

Aerial space treatments have also been used to control tsetse flies *Glossina* spp., although predominantly to minimise the trypanosome disease in cattle, rather than sleeping sickness in man.

Requirements for space treatments

The main objective is to enable spray droplets to remain airborne as long as possible and move slowly downwind, so that insects flying through the cloud of droplets pick up a lethal dose. As many vector species, particularly mosquitoes, typically rest in cryptic habitats not exposed to the drift of airborne droplets, it is important that space spray applications are timed to occur at times of peak vector flight activity, which differs between vectors. The increase in relative velocity between the flying insect (over the resting

Table 3.1 Terminal velocity of small droplets.

Droplet diameter μm	Terminal velocity (m/s)	Time to fall 10 m	Droplet Density (no./cm³)*
1	0.00003	93.7 h	19,120.0
10	0.003	56 m	19.2
20	0.012	14 m	2.38
50	0.075	135 s	0.15
100	0.279	36 s	0.0192

* Applied at 1 litre per hectare.

insect) and the airborne droplet significantly increases the probability of impingement of the droplet upon the insect. An exception to this rule occurs in the case of aerial application of space sprays against tsetse flies. These insects have a very short activity period during daylight hours, when the meteorology may not be favourable for the movement of an aerosol spray cloud to and through the insect's habitat. Instead, space sprays are applied during early morning and early evening hours, when conditions are favourable and the insects are resting in exposed locations on twigs and leaves.

The need to keep droplets airborne dictates that their size is below 30 μm, so that they do not fall by gravity too quickly to the ground (Table 3.1). A number of studies have shown that the optimum drop size to deposit on a mosquito and carry a single lethal dose is between 12-20 μm. Latta et al. (1947) showed that the lowest dose (0.7 mg) of insecticide (DDT) to achieve 50% mortality of *Ae. aegypti* was with 13 μm droplets in air speeds between 3 and 25 km/h. Larger droplets contain in excess of a single lethal dose and thus are wasteful, whereas smaller droplets were not as efficiently collected by the insect, so more droplets were required in the air space and thus a higher dosage. Subsequent studies by Haile et al. (1982) and others have confirmed this early data, so for mosquito control a spray with a volume median diameter (VMD) (Figure 3.3) of 15 μm is regarded as suitable, although thermal fogs using kerosene as the diluent may have a smaller VMD. In contrast, a VMD of about 22 μm was considered optimal for house flies (Yeomans et al., 1949) and 30 μm was recommended for tsetse fly control (Hadaway and Barlow, 1965), although smaller droplets of 20-25 μm will probably be more effective. Larger droplets not only fall faster but there are fewer droplets per litre of spray. With such small droplets, their movement in the environment is very dependent on the meteorological conditions at the time of application. Thus with aerial applications, slightly larger droplets of up to 40 μm may be required initially to use the effect of gravity to pull the droplets downwards from the typical 50-100 m application altitude. Depending on the specific gravity and the volatility of the formulation, the droplets may shrink in size and ideally reach an optimum in the airspace where the vectors are to be found.

(a)

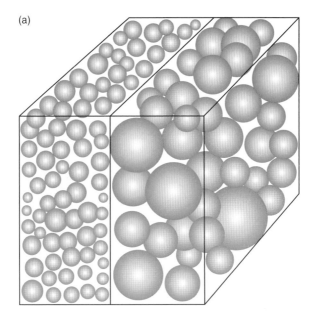

(b)

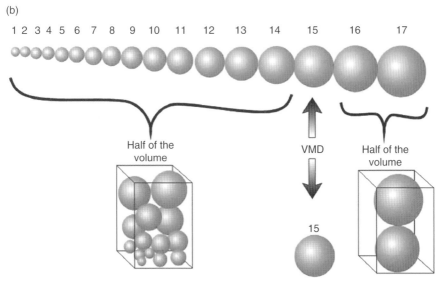

Figure 3.3 Importance of droplet size: (a) Diagram to illustrate volume median diameter (VMD); (b) VMD illustrated to show droplets arranged in size (courtesy of Hans Dobson).

Wind speed inevitably varies but should be less than 4m/s (~15 km/h). A slight air movement is important to allow small droplets to drift through the air, especially to distribute the spray away from the operator or vehicle.

Analyses of droplet spectra from various types of fogging equipment have been reported by Hoffman et al. (2007a, 2008, 2009a). The most suitable way to assess the droplet spectrum of different types of nozzles is by using a laser

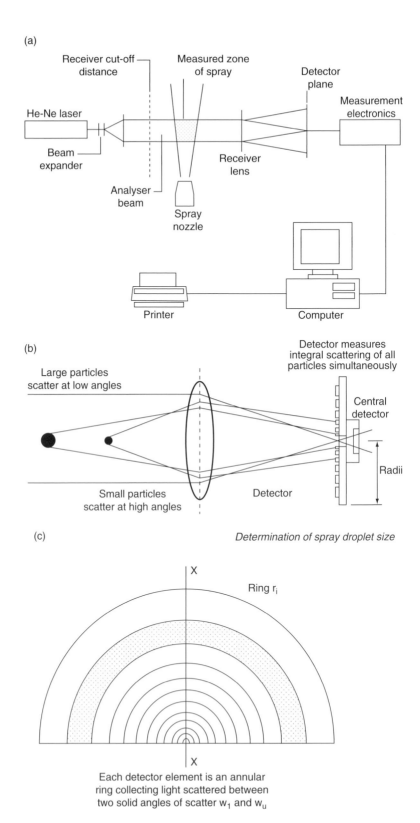

(a)

Receiver cut-off distance

Measured zone of spray

Detector plane

Measurement electronics

He-Ne laser

Beam expander

Analyser beam

Spray nozzle

Receiver lens

Printer

Computer

(b)

Large particles scatter at low angles

Detector measures integral scattering of all particles simultaneously

Central detector

Radii

Small particles scatter at high angles

Detector

(c)

Determination of spray droplet size

X

Ring r_i

X

Each detector element is an annular ring collecting light scattered between two solid angles of scatter w_1 and w_u

Figure 3.4 (a), (b), (c) and (d) Light diffraction analyser to measure spray droplets in flight; (e) measuring spray from a cold fogger.

(d)

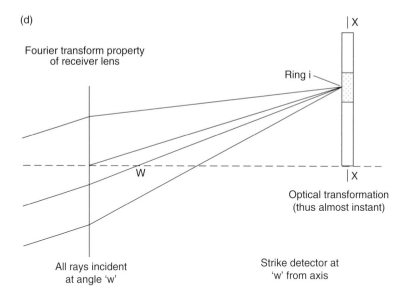

Fourier transform property
of receiver lens

Ring i

| X

| X

W

Optical transformation
(thus almost instant)

All rays incident
at angle 'w'

Strike detector at
'w' from axis

(e)

Figure 3.4 (*Continued*)

light diffraction technique (Figure 3.4a). This equipment is best used indoors, but has been used in the shade of buildings (Figure 3.4e). Another portable measuring device uses a hot wire. Known as the DCIII, it is used to check the droplet spectra of cold fogging equipment, but the sampling unit is very delicate and can easily be damaged by large particles emitted from a nozzle (Figure 3.5).

(a)

(b)

Figure 3.5 (a) DCIII hot wire measuring system for cold fogs; (b) sampling a cold fog.

Equipment for space treatments

The production of the small droplets can be achieved by injecting insecticide in a volatile carrier (e.g. kerosene) into hot exhaust gases in a combustion chamber, a technique referred to as thermal fogging. Traditionally, thermal fogs were used with the insecticide diluted in a large volume of kerosene or similar diluents. Insecticides formulated in water have also been used with adapted equipment but in most cases an adjuvant, for example glycerine, can be added to make the fog more visible. The other criticism of the thermal fog has been the atomisation and dispersal of the diluents, especially oil products into the atmosphere (Mount et al., 1968, Mount, 1998). Nevertheless, a highly visible fog is psychologically more acceptable in some countries. Using caged mosquitoes, Britch et al. (2010) reported that thermal fogging showed greater efficacy than an ultra low volume (ULV) cold fog using malathion at sites in Florida and a hot-arid area of California.

However, the cost of the oil-based diluents and reduced visibility caused by the dense white fog has led to preference for cold fogs, especially when thermal fogs are used in urban areas. Cold fogging uses either a vortex of air or a pressure air jet, which can shear the liquid into the small droplets. Cold fogs can also be achieved with a rotary nozzle operated at a very high velocity typically in excess of 8,000 rpm. Cold fogs can be applied with a minimal volume of liquid, and often are referred to as an ULV treatment. ULV refers only to the volume and should not be confused with other criteria such as droplet size, which is typically in the range of 15-30 μm. However, the number of droplets per unit of volume in a cold fog is far less than when a thermal fog is used with drop sizes typically 3-10 μm diameter, so visibility is not reduced with a cold fogging technique. Thermal fogs are typically applied at a much higher volume than cold fogs, about 3-3.5 litres per hectare for mosquito control versus the 0.02-0.2 litres per hectare typically used in cold fogging. Although thermal fogs are applied at a much higher total volume per unit area, the two methods use similar application rates of active ingredients, the difference in total volume being due to non-active diluents.

Portable equipment

The main type of portable equipment for space treatment is the hand-carried thermal fogger (Figures 3.6, 3.7, 3.8 and 3.9), equipped with a pulse-jet engine, which has been used to treat inside houses, mainly in Southeast Asia and South America and also areas of limited size or inaccessible to vehicles, such as markets, hotel grounds or parks. The advantage of the pulse jet engine is the simplicity of its design and construction. There are no rotating parts and no lubrication is required. The pulse jet engine is started by opening a valve from the fuel tank to the combustion chamber and operating an air piston pump, standard bellows or electric pump to pressurise the fuel tank. Fuel (lead-free petrol) must be metered accurately into the combustion chamber and then mixed with air. The mixture is ignited initially by connecting the spark plug to a capacitor powered by a set of batteries. These machines are sometimes difficult to start, if the fuel mix is not correct. Once the engine ignition is achieved, the hot exhaust gases ignite subsequent charges of fuel and air, so the batteries are no longer needed. The engine should continue to operate with a loud slightly irregular pulsating sound (~80 pulses per second) until fuel is no longer supplied through the carburettor.

The hot exhaust gases pass from the combustion chamber down a long tube, the 'resonator', into which the insecticide is injected (Figure 3.7). The length of the resonator on some versions is smaller to facilitate carrying the equipment through doorways. The temperature at the point of injection can

(a)

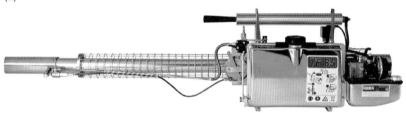

(b)

Figure 3.6 (a) Hand carried thermal fogger; (b) safety mechanism on the fogger (photos: courtesy of Igeba gmbh).

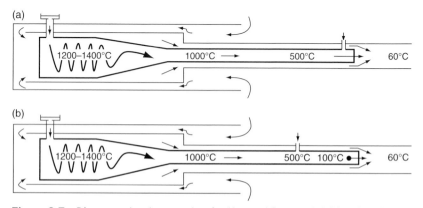

Figure 3.7 Diagrams to show parts of a thermal fogger: (a) Standard hand carried pulse jet fogger for oil based formulations; (b) specialised version for liquids with a flash point above 75°C; (c) close-up of carburetor (from Pulsfog gmbh).

(c)

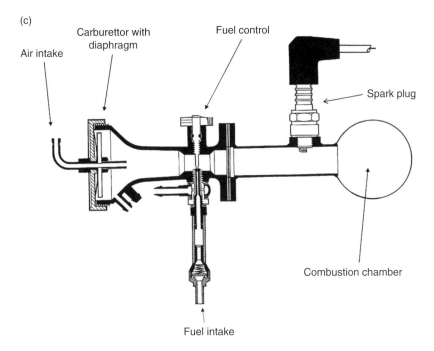

Air intake

Carburettor with diaphragm

Fuel control

Spark plug

Combustion chamber

Fuel intake

Figure 3.7 (*Continued*)

Figure 3.8 Hand carried thermal fogger being used in Subang Jaya, Malaysia following a report of dengue (photo: Chung gait Fee).

be about 500°C, so the liquid is vaporised unless the flow rate is too high. The flow rate should be set using a fixed restrictor. Once the vaporised insecticide in an oil carrier is emitted from the 'resonator' it meets cooler air and condenses to form a dense white cloud, the fog, with droplets generally smaller than 10 μm. When using an insecticide formulation diluted in water, an

(a)

(b)

Figure 3.9 (a) Close-up of hand carried thermal fogger in action (photo: Igeba gmbh); (b) using a hand carried thermal fogger along a street (photo: Pulsfog).

adjuvant, such as mono- or di-ethyleneglycol with an appropriate surfactant, is added to ensure the cloud is visible. It should be noted that water applied through thermal foggers will have a larger drop size, typically 20-40 μm and this can affect efficacy of some treatments, but addition of a glycol reduces the droplet size (Table 3.2). A biological pesticide (*Bacillus thuringiensis*) has been used successfully through this type of fogger, as the particles are exposed to the high temperature for a very short period (0.05-0.1 s), but there are specially designed resonator tubes in which the temperature can be lowered prior to the injection of the insecticide (Figure 3.12b).

Table 3.2 Preparation of insecticide liquid for fogging (volumes in litres) (adapted from Information supplied in Pulsfog leaflet).

Desired droplet spectrum	<10 μm	<25 μm	<50 μm
Composition of fog liquid	*100% oil*	*50% oil or glycol in water*	*25% oil or glycol in water*
Spray oil with emulsifier	3.0*	1.5	0.75
Glycol		1.5	0.75
Wetting agent		0.06	0.03
Water		1.5	2.25
Total	3.0	3.06	3.03

* Emulsifier is not mixed with oil when only oil is used as the diluent.

The equipment should be fitted with a safety valve that cuts off the flow of liquid if the engine ceases to operate (Figure 3.6b). The operator should also have a fire extinguisher readily available should a flame be emitted from the resonator. The equipment should not weigh more than 20 kg with full fuel and insecticide tanks and be easily carried with a shoulder strap. The resonator should be directed behind the operator so that he walks away from the fog, but because the droplets are so small, many being less than 10 μm diameter and could be inhaled, the operator must wear respiratory protective equipment (RPE). Operators should also protect their ears due to the intense noise level. These factors necessitate that only those who have been well-trained and fully understand their operation and routine maintenance should be entrusted with thermal foggers.

Sufficient fog is usually emitted to treat 200 m³/min that can cover an area of 3 hectares in an hour. A fog should be applied only when the wind speed is less than 4m/s, even when treating inside buildings, as a fog can be sucked from an enclosed building through the smallest gaps. In treating a house for vector control, it is important to keep the house closed for a period after treatment and then allow sufficient time, at least 30 minutes, for ventilation before allowing people to re-enter. Fogs are sometimes applied around refuse dumps and large accumulations of old tyres (Figure 3.10), although such treatments only affect the insects flying in the area and may need to be repeated on several consecutive days to reduce migration of insects emerging from within the dump.

A different type of thermal fogger is fitted with a 1–3 hp 2-cycle engine driving a 'friction plate' inside the insecticide tank, which pre-heats the insecticide liquid and pumps the liquid to the engine exhaust. Although it operates at a lower temperature, it is not suitable for applying a biological insecticide as the friction plate damages spores.

Some small portable cold foggers with small engine with gearbox and air turbine are available (Hoffman et al., 2007a). but small cold foggers with an electric motor and a timer to operate at a pre-set time have not been used in vector control. Knapsack cold foggers, with a vortical nozzle (Figure 3.11) that

Figure 3.10 Thermal fogger being used to treat old tyres where mosquito larvae can develop (photo: Igeba gmbh).

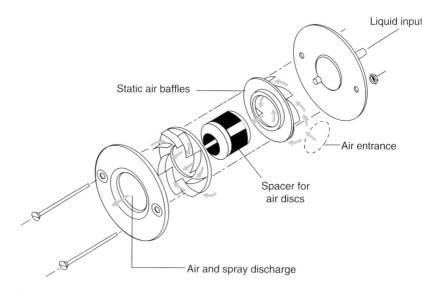

Figure 3.11 Cutaway of a cold fogger vortical nozzle.

relies on a vortex of air to shear liquid into the small droplets without heat, are commercially available. These have a 2-cycle engine to drive a compressor to provide the air flow to the nozzle. Thermal foggers have been preferred to portable equipment, because of the visibility of the fog.

Mist treatments

Mist blowers, which produce a spray with a VMD larger than 50 μm, and project the droplets in an air stream, have also been used (see Barrier treatments in Chapter 6). With these larger droplets, but still defined as a fine spray, a more residual deposit can be achieved, depending on the dosage applied. Experiments with motorised knapsack mist blowers were conducted in Nigeria, applying technical malathion in a village with 582 houses near Enugu. Approximately 160 ml was applied per house, making a total of 90 litres used indoors and a further 62 litres were applied outdoors. There were two applications with a week interval between. Most of the spray indoors was deposited on the floor with irregular distribution on walls, depending on where the nozzle was directed. Captures of mosquitoes remained low in the treated village compared with an unsprayed village for at least 20 weeks, and this was attributed to the larger droplet size than in a space treatment, giving a residual effect (Bown et al., 1981). An improved residual effect achieved with a mist compared with fogging can also help reduce populations of dengue vectors when mists are applied inside dwellings.

Mist treatments have also been applied around buildings to reduce infestations of flies, which can spread infections such as trachoma of the eye caused by *Chlamydia trachomatis* and diarrhoea (Emerson et al., 1999). Recently there has been renewed interest in mist applications as barrier treatments to vegetation around housing areas, to kill mosquitoes before they enter houses (Trout et al., 2007). Penetration of spray into vegetation is improved with a higher air velocity (Hoffmann et al., 2009b). In a test in Florida, one pyrethroid spray suppressed a mosquito population below an annoyance action threshold of 25 mosquitoes per night in carbon dioxide baited light traps for 5 weeks (Cilek, 2008). In a small village, Matthews et al. (2009) combined the mist treatment with screening of houses to reduce entry of mosquitoes into the houses. Mist blowers have also been used successfully to apply larvicides to water surfaces in small container breeding areas in marsh lands and around residential areas – a technique referred to as broadcast larviciding.

The knapsack mist blower has a 2-cycle engine to drive a centrifugal fan. The air passes down a flexible duct at the end of which is a shear or rotary nozzle. There should be a fixed metering restrictor to control the optimum flow of spray liquid to the nozzle. Too high a flow rate results in poor atomisation and large droplets that do not remain in the air stream. The air velocity at the nozzle is about 60–75 m/s, but dissipates very quickly. The engine on this equipment must be used at the recommended optimum speed and not allowed to 'idle', otherwise the engine will be damaged by inadequate lubrication. The engine may be difficult to start when fuel has been left in it. Petrol will vaporise and leave an oil deposit on the spark plug, so when a day's operation is complete, it is better to stop the engine by switching off the fuel supply, allowing the engine to run until starved of fuel.

Small rotary nozzles (spinning disc) with a fan have been considered for treating around latrines, to provide a residual mist application where vectors may rest. Indoor space spraying has also been investigated with a ULV cold fog applied, either through the front door for one minute or into each room for 15 seconds, using a CS formulation of λ-cyhalothrin. Both methods of application provided significant control over 3 weeks, based on adult *Aedes aegypti* house collections (Perich et al., 2003).

Vehicle mounted equipment

Larger truck mounted thermal foggers are also manufactured and have been used in vector control. These may be similar to the hand carried equipment with a large pulse jet engine, but others have a burner and blower unit (Figure 3.12). However, their use is not generally recommended, due to the large volume of diluents needed and the potential for a dense fog to cause a traffic accident. The emphasis is therefore for vehicle mounted equipment to produce a cold fog using a low pressure (vortical nozzle) (Figure 3.11), high speed air shear or rotary nozzle (Figure 3.13). The majority of vehicle mounted cold foggers have a gasoline engine to drive a high volume air blower. Approximately 6 m³ of air is delivered to one or more vortical nozzles at low pressure (50 kPa). On older equipment, the insecticide tank was pressurised, but now the equipment usually has a positive displacement pump to control the flow from the insecticide tank to the nozzles, as the output can be regulated in relation to the vehicle speed, or stopped when necessary if the vehicle stops. The latest equipment has controls in the vehicle cab and can be linked to a global positioning system, so that there is a record of the route taken, the time and the amount of insecticide applied. This type and equipment with a high pressure shear nozzle is very noisy, so some authorities have used a much quieter design with an electrically driven rotary nozzle. This nozzle rotates at over 25,000 revolutions per minute.

The route for a vehicle mounted fogger will depend on the street layout (Figure 3.14). A track spacing of 50 m is advised, but this may not always be possible. In the USA, the label specifies a track spacing of 300 feet (~100 m) for the purposes of flow rate calibrations, since this is the average urban block size (street spacing). However, track spacing should be based on the average street layout of the target town/city/country and the equipment calibration must be based on that and not on any other factor. The actual swath achieved will depend on wind speed and the impact of buildings and vegetation on the spray cloud. Ideally the cloud of droplets is directed upwards and slightly downwind behind the vehicle, so that a cross wind will divert droplets at different heights above the road and disperse them over as wide an area as possible, downwind of the vehicle. In some situations, the nozzle should be mounted high on a vertical duct, so that spray droplets are released above the height of any wall alongside the road. As with the portable equipment, the cloud of droplets will be dispersed too rapidly if there is too much wind, so the wind speed should be less than 4 m/s. If the droplets are released outside buildings over hot ground, local thermal convective currents of warm air can lift the droplets, thus in some areas applications are ideally completed very early in the morning if the vector is active at this time. In urban areas, when

(a)

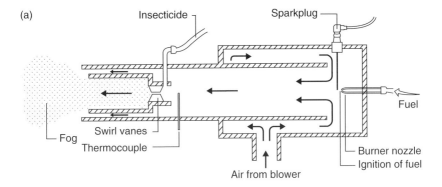

(b)

Figure 3.12 (a) Diagram of a vehicle mounted thermal fogger with air blower; (b) truck mounted thermal fogger in Egypt.

hot air is expelled outside a building by an air-conditioning unit, small droplets will be lifted up and over the building. All windows and doors should be open during an application, but relatively few small droplets will penetrate into a building. Similarly in vegetated areas, droplets do not penetrate dead space within a bush canopy (Lothrop et al., 2007). In built-up areas, the dispersal will be very much affected by air movements deflected by buildings.

With spray droplets moving downwind (Figures 3.15, 3.16 and 3.17), the vehicle should commence a treatment on the downwind edge of that area and move along parallel tracks progressively upwind, so that the fog is carried downwind from the track. Roads with a dead end should be treated only by starting at the end, and moving towards the exit. Treatment should continue for at least 300 m upwind of an area to reduce re-invasion from a neighbouring area, but local conditions and potential larval breeding sites need to be considered when planning a space treatment.

A space treatment will only kill the adult mosquitoes exposed to the airborne droplets, so as larvae continue to develop and pupate and more adults emerge after a treatment, it is necessary to apply a minimum of 3 sequential space treatments up to 6–7 days apart to reduce the spread of a viral infection such as dengue, as adults emerging after the initial spraying can be infective due to trans-ovarial transmission of the virus by the mosquito (Figure 3.13e). Esu et al. (2011) concluded that there was no clear evidence for recommending peridomestic space spraying, but only a few studies reviewed by them included

(a)

(b)

Figure 3.13 (a) Truck mounted cold fogger being operated in Subang Jaya, Malaysia (photo: Chung gait Fee); (b) truck mounted cold fogger used in Florida (photo: Mark Latham); (c) twin spray cold fogger (photo: John Clayton); (d) cold fogger with rotary atomiser (photo: Graham Parker); (e) timing of sequential space sprays to control mosquito populations when human cases of dengue are detected at the start of an outbreak of the disease.

(c)

(d)

(e)

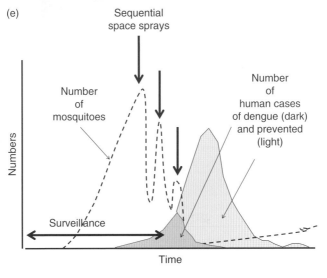

Figure 3.13 (*Continued*)

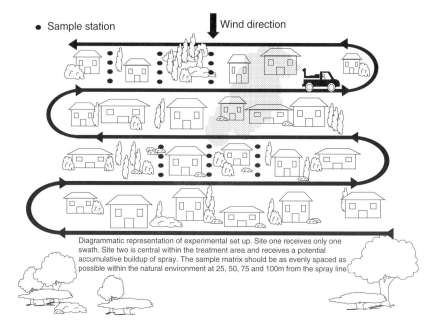

Figure 3.14 Diagram to show passage of a truck mounted fogger through a housing area. In many areas there may be "no-through" roads so careful planning of a route is needed. Vehicles may have to drive down a "no-through" road and start to fog at the end and move back to the entry of the road. The route should also optimise use of the wind direction to spread the fog away from the road and through the housing area. (Reproduced with permission from WHO.)

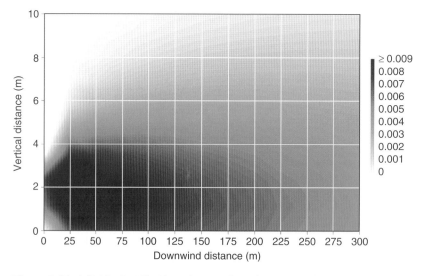

Figure 3.15 LIDAR visualisation of spray from truck mounted cold fogger (courtesy of Bob Mickle). Most spray in the darkest area.

(a)

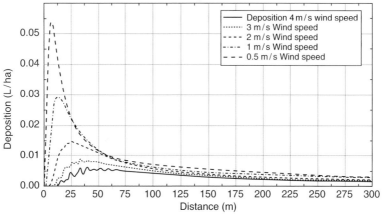

(b)

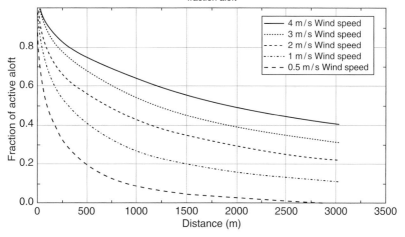

Figure 3.16 (a) Deposition on ground downwind of a cold fogger – droplets released at 2 m; (b) fraction of active spray still airborne at different distances downwind (courtesy of Bob Mickle).

appropriate sequential area-wide treatments. Nevertheless, a major reduction in the adult mosquito population can be obtained by fogging correctly when there is an extensive emergency to control a dengue outbreak and this gives more time to implement other control measures, particularly reducing larval breeding sites in urban areas.

One of the insecticides used for this type of treatment has been technical malathion, which was applied undiluted at ULV. Care has to be taken to ensure that the droplets are the recommended size, otherwise vehicle paint could be affected by the spray. The trend has been to apply specially formulated low volatility formulations diluted in water to minimise any effects on foliage of

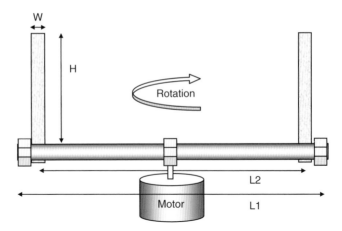

Figure 3.17 Rotary sampler for collecting small airborne droplets (H = 75 mm; L1 = 22 cm; L2 = 18 cm; W = 3 mm) (reproduced with permission from WHO).

roadside plants. More recently, other insecticides, mostly in the synthetic pyrethroid family, have been preferred to avoid the odour of the spray.

A number of urban municipalities also now use water miscible formulations, some with anti-evaporant properties, to apply as a cold fog and often relying on some limited evaporation of water to reduce spray particle size to the optimal. Water is preferred to oil as a carrier and diluents, as it does not stain surfaces and is less expensive.

Aerial application

In emergency situations, aerial application may be used when very large areas require rapid treatment. Some aircraft have been converted to apply a thermal fog by injecting the insecticide into an adapted engine exhaust, although these are not widely used. Until recently it was believed that the shearing effect due to the high speed of an aircraft would allow hydraulic nozzles to be used and still achieve an optimum droplet size with a VMD of less than 40 µm. However, studies in the late 1990s demonstrated true VMDs from hydraulic nozzles on aircraft to be between 60 and 100 µm (Hornby et al., 2006). The trend is now towards using electrically driven high speed rotary nozzles, the rotational speed of which can be controlled electronically. These nozzles have been demonstrated to produce droplets with a VMD of between 25 and 40 µm. Early large-scale aerial trials demonstrated effective control of *Ae. aegypti* in Thailand by applying malathion (Lofgren et al., 1970). In the southern USA (particularly Florida, Louisiana and Texas) aerial adulticide sprays are commonly applied by local government programmes against large populations of nuisance flood water mosquito species, such as *Ae. taeniorhynchus*. Typically this amounts to 2 million to 4 million hectares per year in Florida alone (Mark Latham, pers. comm.). In addition to these 'routine'

(a)

(b)

(c)

Figure 3.18 (a) US aircraft spraying over New Orleans, September 2005, following Hurricane Katrina; (b) helicopter spraying against adult mosquitoes in Florida; (c) fixed wing aircraft spraying at night; (d) kitoon to raise meteorological equipment to measure wind speed and other variables above area to be sprayed (photos: Mark Latham).

(d)

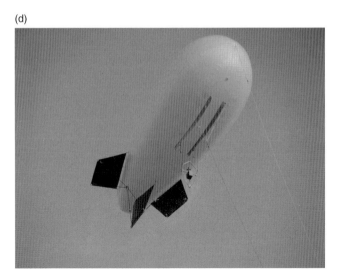

Figure 3.18 *(Continued)*

programmes, emergency aerial applications are utilised following damage by hurricanes (Figure 3.18) to control *Ae. atlanitcus/tormentor*, the West Nile virus vector *Culex nigripalpus*, *Ae. vexans*, and the floodwater pest mosquito *Psorophora columbiae* (Simpson, 2006). In one example, a 90% decline in mosquito abundance was observed post-treatment after spraying over 100,000 ha with naled (Breidenbaugh et al., 2008). The US Air Force (USAF) has developed a modular aerial spray system (MASS) for use with the C-130H aircraft. The advantage is that it can be loaded on the airframe in less than 1 hour and has a 7,571-litre capacity (Breidenbaugh and Haagsma, 2008). Flat fan nozzles were normally mounted on the wings, but a novel boom configuration on the fuselage with TeeJet 8005 nozzles directed downwards has been developed for the C-130H aircraft. Unlike typical mosquito control aircraft, whose flying speeds of 200–270 km/hr cannot sufficiently atomise sprays from hydraulic nozzles to a VMD below 50 µm (Hornby et al., 2006), the 370 km/hr flight speed of the C-130H provides sufficient energy through shearing to atomise the spray from hydraulic nozzles to an optimum VMD of less than 40 µm. In recent aerial spray trials using the C-130H flying at 370.4 km/h and a height of 46 m spraying naled, 100% mortality of caged mosquitoes was achieved 639 m downwind with a single pass and generally 90% control was attained up to 1,500 m downwind (Breidenbaugh et al., 2009). Droplets were recorded as 44 µm at 213 m and 11 µm at 2130 m downwind. In area-wide control operations using the C-130H applying naled over Paris Island in South Carolina, biting midge populations were reduced by 86% on average after 7 days. Although the majority of the droplets were much larger than the 12–20 µm originally considered optimum for space treatments, other factors such as height of release (50 m vs 5 m for truck mounted equipment) become important. As such, the larger droplets

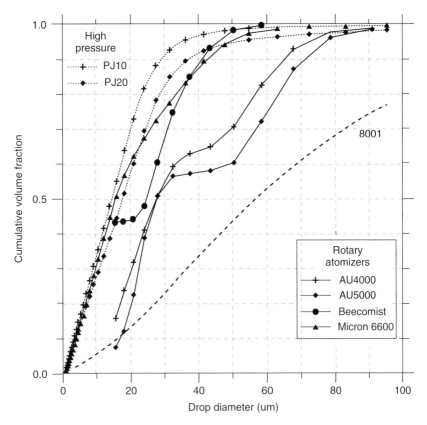

Figure 3.19 Aerial application of adulticides, showing droplet spectra from high pressure and rotary nozzles in comparison with hydraulic 8001 nozzle operated at 80 psi (data courtesy of Bob Mickle).

contained in a spray with a VMD of about 40 μm are able to fall to the target zone and achieve a high mortality of mosquitoes.

To optimise sprays with a VMD of less than 25 μm, two types of nozzles that can be mounted on aircraft have been shown to be suitable – high pressure nozzles, although some of these are prone to clogging, and rotary atomisers (Figure 3.19). The latter have been improved by increasing their rotational speed and re-designing the screen through which the spray is emitted. With electrically operated rotary atomisers, droplet size is independent of the aircraft speed. On the Micronair AU6600, VMDs ranged from 18–24 μm as flow rate increased from 0.5–1.5 l/min. while the input current increased from 10–13A (350 W). The number of atomisers per aircraft would depend on the maximum power available and the flow rate required. A total volume of 9 l/min can be applied with 6 AU6600 atomisers, if 90 A is available on aircraft flying at 240 km/h treating 100-m swaths. On slower helicopters (80 km/h), two atomisers could provide 3 l/min, unless a wider swath is required when additional atomisers could be fitted.

High pressure hydraulic nozzles operated at 1,500–3,000 psi with flow rates ranging from 0.4–1.3 l/min can achieve a suitable droplet spectrum. On fixed wing

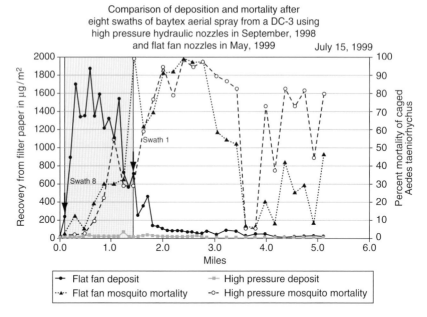

Figure 3.20 Downwind deposition and mortality of mosquitoes following aerial application using hydraulic high pressure and flat fan nozzles with fenthion in May 1999. The spray started with swath 1 and continued upwind until swath 8 had been applied. Arrows show position of first and last swath.

aircraft, a minimum of 10 nozzles operated at 1,500 psi might be needed and with an electric pump delivering 8 l/min, the power requirement would be 90 A.

A high pressure hydraulic nozzle (1/8 MIS) operating at 3,000 psi (Robinson, 2001) evaluated in a wind tunnel with a wind velocity 277 km/h, generated aerosol droplets within a 20–30 μm VMD range, as shown using a DCIII hot-wire measuring system (Figure 3.5). In a field tests with the prototype, single swaths indicated that 50% of the spray had deposited from flat fan nozzles within 0.4 km and 100% by 5 km., whereas the high pressure nozzle deposited 7 and 47% at the same respective distances. The greater deposition with the flat fan nozzle applying 38 ml/ha fenthion resulted in 80–90 % mortality of terrestrial fiddler crabs, while there was no mortality with the smaller droplets remaining airborne longer (Dukes et al., 2001) (Figure 3.20). Modelling showed the distribution of droplets of specific sizes (Figure 3.21). Similarly in another investigation, a high-pressure nozzle system substantially reduced environmental insecticide contamination and decreased bee mortality (Zhong et al., 2004). With bees in their hives at night, spraying against mosquitoes late in the evening/night with a low dose of insecticide in small droplets leaving no residual deposit, the risk of bee mortality is extremely low. Furthermore, application using high-pressure hollow cone nozzles needed half the amount of insecticide to control mosquitoes compared to flat-fan nozzles (Dukes et al., 2004). Recent advances in high pressure systems include the use of impingement nozzles. These nozzles provide a greater degree of atomisation over regular high pressure hydraulic nozzles, as the

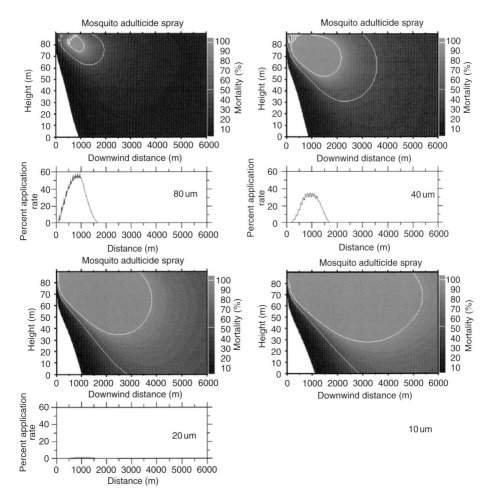

Figure 3.21 Data from model to illustrate downwind movement with four different droplet sizes. Darkest area has fewest droplets thus more deposition within 1500m downwind with the larger droplets. (data courtesy of Bob Mickle).

high pressure stream exiting the fine nozzle orifice encounters a 'J Pin' causing a secondary atomisation through the shattering effect. Impingement nozzles provide a similar droplet spectrum to regular high pressure nozzles but at less than half the operating pressure (500–1,000 psi vs. 1,500–2,500 psi).

As an alternative to flat-fan nozzles and high pressure systems, rotary atomisers, such as Micronair atomisers, have been used in mosquito control investigations (Figure 3.22). Elnaiem et al. (2008) sprayed pyrethrin in an urban/suburban area in California and showed that *Culex* spp. abundance and biting rates were reduced and decreased the risk of West Nile virus infection. The addition of oil to reduce the effect of low humidity on droplets has also been studied (Lothrop et al., 2007). Rotary atomisers have also been used in tsetse control operations in Africa (Lee et al., 1975; Allsopp, 1990). Brown et al. (2003) point out that proper flight line positioning is critical to achieve maximum efficacy when spraying in cross winds using small droplets spray strategies.

(a)

(b)

(c)

Figure 3.22 Micronair rotary atomiser used in mosquito adulticiding;
(a) Micronair Spray Pod with AU4000 rotary atomiser on BN Islander aircraft;
(b) Micronair AU4000 atomiser on fixed wing aircraft; (c) Micronair AU6539
electric rotary atomiser on Hughes 500 helicopter in USA, showing the spray
tank behind the pilot (photos: Micronair).

Insecticides

Although the dosage needed for a space treatment is substantially less than for residual sprays (typically only 1% of the residual amount), it is essential that only insecticides with a low mammalian toxicity at the required doses are applied, particularly where applicators and members of the public may be exposed to a concentrated spray cloud close to the equipment/nozzle (applications by hand-held or truck-mounted equipment). A series of sequential applications are applied as the effect is not residual, so mosquitoes emerging from pupae need to be controlled by subsequent sprays before they oviposit. The main insecticides used in space treatments have been the organophosphates, malathion and fenitrothion and various pyrethroids, including deltamethrin, λ-cyhalothrin, permethrin and resmethrin (Table 3.3). The organophosphate naled is an example of an insecticide with a relatively higher mammalian toxicity that is appropriate for aerial space spray applications, but not for hand-held or truck-mounted applications. The toxicity and irritation/corrosiveness of the concentrated spray cloud is mitigated by significant dilution occurring before it reaches the ground. Formulations of naled are widely used in aerial applications in the USA, based on the environmental benefits of an extremely short half-life and the application benefits of a high specific gravity (1.8), which aids in the fall velocity of the extremely small droplets.

Formulations suitable for space treatments include specialised fog and UL formulations. Low volatility of the spray is required where optimum droplet size can be achieved. In some case such as aerial application, when the aircraft has to fly high to avoid trees and other obstructions, an initially larger droplet size can be applied allowing the volatility of the formulation to allow the droplet to

Table 3.3   Insecticides recommended for space treatments by WHO. The list does not include some insecticides recommended in certain parts of the world.

Insecticide	Type	Dose (g a.i./ha) Cold aerosol/fog	Dose (g a.i./ha) Thermal fog
Fenitrothion	OP	250–300	250–300
Malathion	OP	112–600	500–600
Cyfluthrin	Pyrethroid	1–2	1–2
Cypermethrin	Pyrethroid	1–3	–
Cyphenothrin	Pyrethroid	2–5	5–10
D,d-trans-Cyphenothrin	Pyrethroid	1–2	2.5–5
Deltamethrin	Pyrethroid	0.5–1.0	0.5–1.0
D-Phenothrin	Pyrethroid	5–20	–
Etophenprox	Pyrethroid	10–20	10–20
λ-Cyhalothrin	Pyrethroid	1.0	1.0
Permethrin	Pyrethroid	5	10
Resmethrin	Pyrethroid	2–4	4

shrink as it descends. In contrast to indoor residual spraying, the use of particulate formulations, such as wettable powders, suspension concentrates and water-dispersible granules should not be used for space treatments.

The efficacy of insecticides as space sprays can be evaluated in the laboratory using a small wind tunnel in which caged mosquitoes are exposed to 15 + 2 μm droplets VMD in the air flow (Figure 3.23) (WHO, 2009). Hoffman et al. (2008) have shown that the type of mesh can affect the results of studies with caged mosquitoes by filtering out droplets. If mosquitoes are not transferred within

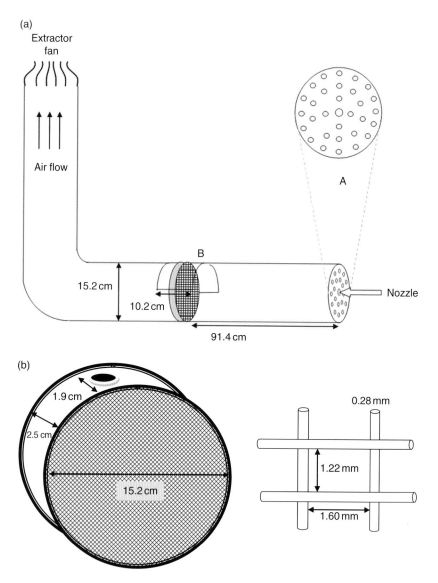

Figure 3.23 (a) Wind tunnel for assessing effect of aerial sprays; (b) cage to be inserted into the wind tunnel; (c) insertion of cage into the wind tunnel (photos: J. Bonds).

(c)

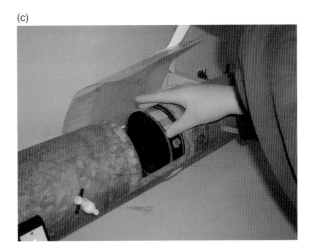

Figure 3.23 *(Continued)*

about 15 minutes to clean cages, the mortality is increased by exposure to the deposits on the mesh (Barber et al., 2006). When using a wind tunnel test, Bonds et al. (2010) compared different mesh sizes and showed that mortality was higher with a more open mesh. Tests also confirmed that spray collected on the mesh caused similar mortality but not as quickly as the airborne droplets. The positioning of sentinel cages in the field during trials to assess the efficacy of sprays is facilitated by having suitable supports (Clayson et al., 2010).

The risk of applying insecticides as space sprays has been considered in the USA, where the incidence of West Nile fever has emphasised the importance of controlling the disease vectors (Peterson et al., 2006). Space sprays are a major component in the IPM approach of most mosquito control programs in the USA. For example, the average annual acreage treated by space spray applications by mosquito control programmes in Florida over the past 10 years is 2–4 million hectares by aircraft and 6–10 million hectares by trucks. WHO has published generic guidelines on use of space sprays (WHO, 2010).

Planning

Space treatments are generally in response to an emergency when a disease outbreak occurs, or on a smaller scale, members of the public require urgent attention to a local problem. When urgent action is needed, it is essential that there has been sufficient forecasting of needs, stocking sufficient insecticides and preparation of equipment so that an immediate response is possible (Box 3.1). Thus prior to the season most favourable for outbreaks of disease or nuisance due to increased populations, detailed mapping of an area needs to be undertaken, equipment checked and maintained and supplies of all the materials needed ordered well in advance of their intended use. In Africa, a control programme immediately prior to the wet season will require initial

Box 3.1   A Check List

Target vector

Species; habitat; survey data/population density

Adults or larvae?

Area requiring treatment?

Equipment availability

Sprays and/or granule application

Ground equipment (portable, vehicle mounted)

Aerial application (fixed wing aircraft/helicopter availability, application equipment?)

Timing/seasonality of problem (*Rainfall data*)

Meteorology conditions affecting treatments?

Wind velocity/temperature inversions

planning not less than 6 months before the start of control operations. Training of staff at this time is needed, so everyone involved knows exactly their role in the control operations.

Assessment of space sprays

Traditionally, the efficacy of sprays during experimental trials has been evaluated by placing caged mosquitoes at different distances downwind and simultaneously either sampling the spray deposit on ground targets or on rotating samplers. The difficulty with caged mosquitoes is that the material used in the cage will collect spray, so the efficacy reported is strictly due to the caged mosquitoes contacting the deposit on the mesh rather than as free-flying insects (Boobar et al., 1988). The risk is the assumption that larger droplets are more effective as more insecticide is collected on the cage. The main method for evaluating efficacy of large area operational applications is using traps, usually with the addition of dry ice where it is available (Sudia and Chamberlain, 1962; Reiter, 1983; Service, 1993), or by using a battery powered back pack aspirator (Scott et al., 2000; Vazquez-Prokopek et al., 2009).

Bonds et al. (2009) have reported that an FLB rotary sampler (Figure 3.17) with 3-mm wide slides on an 18-cm arm rotating at 600 rpm (5.6 m/s slide velocity) is more effective for sampling small droplets below 30 μm than the previous standard ('Hock Spinner'), which utilised microscope slides 25 mm wide on a 13-cm arm rotating at 450 rpm (3.1 m/s slide velocity). This is because the collection efficiency of very small aerosol droplets is improved on a very narrow slide travelling at a faster speed. Similarly, Johnstone et al. (1990) used 6.3-mm-wide slides on rotary samplers to assess aerial sprays applied in tsetse fly (*Glossina* spp.) control.

In Thailand in the Kamphaeng Phet area, Koenraadt, et al. (2007) applied a cold fog indoors and outdoors using a Colt ULV aerosol generator. The results clearly demonstrated the lack of persistence as *Aedes* populations were at least 50% of pre-spray numbers within a week, especially at the edge of a treated area. Spraying outside had no beneficial effect. Apart from breeding sites within a treated area, mosquitoes could rapidly migrate in from other areas or from water sources within the sprayed area.

In experiments in California, cage mortality agreed well with droplet density measured on rotating Teflon™ coated slides, and that mortality was affected by distance from the truck, obstructions such as trees, low wind speed, and the failure of droplets to penetrate vegetation. In these experiments, changes in abundance of mosquitoes was also measured using dry-ice baited traps and the recapture patterns of marked females released near traps in treated and unsprayed areas. Overall, the relative abundance of *Culex tarsalis* always declined after ULV applications (Lothrop et al., 2007a).

Confirmation that the optimum droplet size for mosquito control was obtained when data from 76 field trials analysed the mortality of caged *Ae. taeniorhychus* adult mosquitoes situated in a vegetated habitat using a 4-nozzle vehicle mounted cold fogger applying permethrin. The air output was regulated to maintain droplet size (VMD) at 7, 15 and 26 µm. The most effective droplets were 15 µm, while the smallest droplet size gave the lowest mortality in samples up to 150 m downwind. The 26 µm droplets were effective from 30–100 m downwind (Curtis and Beidler, 1996). Ground deposits would be expected to decline as the larger droplets sedimented, but in experiments reported by Moore et al. (1993), ground deposits decreased at the furthest sample (91 m) only during one trial.

Comparing open country with areas with more vegetation, tests in Florida showed that 3.25 times more spray was detected in open areas and this resulted in a 10-fold increase in mortality (Rathburn and Dukes, 1989). Barber et al. (2007) confirmed that control was better in open areas.

While assessing the efficacy of treatments, Moore et al. (1993) also measured the extent to which two stationary bystanders (at 7.6 and 15.2 m) and a jogger moving in the same direction as the vehicle but 1.5 m from the spray path, were exposed to technical malathion (58.5 g/ha) applied in the evening between 1715 and 1915 hours, using a vehicle mounted ULV cold fogger (Leco HD) travelling at 16 km/h. The deposit measured on the jogger with the least protection showed that an adult male (Table 3.4) weighing 70 kg would need to be exposed to over 35,000 similar applications to accumulate the reported LD_{50}.

Methods for assessing the efficacy of space sprays applied with ground equipment have been recommended by WHO (2009).

Naled has been used widely in the USA for adulticiding, due to its short persistence. Brown et al. (2006) assessed the impact of low application rates of naled (Dibrom 14 concentrate), using aerial application with a Micronair AU4000 rotary atomiser (Figure 3.22). They reported that bio-assays with caged adult *Anopheles quadrimaculatus* in open grassland indicated that a

Table 3.4 Amount of malathion (mg) detected on the persons exposed to a ULV space spray.

Person	Distance	Arms not protected Short trousers	Arms not protected Long trousers
Jogger	1.5 m	7.80	6.13
Bystander	7.6 m	2.30	1.36
Bystander	15.2 m	1.78	1.18

dosage of 18 ml/ha or less should be avoided and that a dose greater than 36 ml/ha may be required for adequate control in canopied habitats. Using a truck mounted cold fogger to apply naled at 22.4 g/ha at 44 ml/min, Schleier and Peterson (2010) found no quantifiable air concentrations of naled between 1 and 12 hours after application in two separate trials. In these trials, similar results were obtained when applying permethrin.

Bio-monitoring of urine showed that pilot exposure to pyrethrin was extremely low during aerial applications of a ULV application for adult mosquito control. The highest exposure was equivalent to a dosage of 0.03 mg pyrethrin/kg/day, which represents approximately 1/2,800,000th of the lowest observed adverse effect level and 1/1,000th of the acceptable daily intake for pyrethrin. Thus aerial application of ULV pyrethrin did not result in undue exposure to a trained and certified pilot (Gerry et al., 2005). A subsequent study during an aerial spray control programme to reduce *Culex* spp. populations following detection of West Nile virus, confirmed that exposure to the pyrethrin and piperonyl butoxide spray was below thresholds set by the Environmental Protection Agency. The benefit of reducing the risk of virus infection far outweighed the public health risk due to using the insecticide (Macedo et al., 2010).

Monitoring

Monitoring adult mosquito activity is important, not only to check on the effectiveness of control operations, but also to determine the variations in mosquito populations and determine or forecast when numbers will exceed an acceptable threshold and further control operations are needed. Differences in the behaviour of vector species requires adjustment of sampling procedures, choice of trap design and their position within the area in which people need to be protected. In one assessment of two types of traps, the gravid trap caught significantly more than a CDC light trap baited with CO_2 placed under houses, although more species were caught in the light trap (White et al., 2009).

Conclusion

Space treatments are an effective means of rapidly reducing high populations of both nuisance and disease vector mosquitoes and other flying insects of public health importance. The technique depends primarily on the scale of the operation and the weather conditions at the time of treatment, but also includes considerations related to the species, their behaviour, the target habitat, the cultural environment and the level/geographical extent of the disease outbreak (Box 3.1).

References

Allsopp, R. (1990) A practical guide to aerial spraying for the control of tsetse flies (*Glossina* spp.) *Aerial Spraying Research and Development Project Final Report*, vol 2. Natural Resources Institute, Chatham.

Anon (2010a) *Sofia News Agency*, 1 September 2010.

Anon (2010b) *Macedonian International News Agency*, 28 August 2010.

Barber, J. A. S., Greer, M. and Coughlin, J. (2006) The effect of pesticide residue on caged mosquito bioassays. *Journal of the American Mosquito Control Association* **22**: 469-47.

Barber, J. A. S., Greer, M. and Coughlin, J. (2007) Field tests of malathion and permethrin applied via a truck-mounted cold fogger to both open and vegetated habitats. *Journal of the American Mosquito Control Association* **23**: 55-9.

Barzon, L., Squarzon, L., Cattai, M., et al. (2009) West Nile Virus infection in Veneto region, Italy, 2008-2009. *Eurosurveillance* **14(31)**: article 6.

Bonds, J. A. S., Greer, M., Fritz, B. K. and Hoffmann, W. C. (2009) Aerosol sampling: comparison of two rotating impactors for field droplet sizing and volumetric measurements. *Journal of the American Mosquito Control Association* **25**: 474-9.

Bonds, J. A. S., Greer, M., Coughlin, J. and Patrl, V. (2010) Caged mosquito bioassay: studies on cage exposure pathways, effects of mesh on pesticide filtration and mosquito containment. *Journal of the American Mosquito Control Association* **26**: 50-6.

Boobar, L. R., Dobson, S. E., Perich, M. J., Derby, W. M. and Nelson, J. A. (1988) Effects of screen materials on droplet size frequency distribution of aerosols entering sentinel mosquito tubes. *Medical and Veterinary Entomology* **2**: 279-384.

Bown, D. N., Knutsen, A. B., Chukwuma, F. O., et al. (1981) Indoor and outdoor ULV applications of malathion for extended control of *Anopheles* and *Aedes* species in wooded rural communities in eastern Nigeria. *Mosquito News* **41**: 136-42.

Breidenbaugh, M. and Haagsma, K. (2008) The US Air Force Aerial Spray Unit: a history of large area disease vector control operations, WWII through Katrina. *Army Medical Department Journal* April-June, 54-61.

Breidenbaugh, M. S., Haagsma, K. A., Wojcik, G. M. and de Szalay, F. A. (2009) Efficacy of aerial spray applications using fuselage booms on Air Force C-130H aircraft against mosquitoes and biting midges *Journal of the American Mosquito Control Association* **25**: 467-73.

Breidenbaugh, M. S., Haagsma, K. A., Walker W. W. and Sanders D. M. (2008) Post-Hurricane Rita mosquito surveillance and the efficacy of air force aerial applications for mosquito control in east Texas. *Journal of the American Mosquito Control Association* **24**: 327-30.

Britch, S. C., Linthicum, K. J., Wynn, W. W., et al. (2010) Evaluation of ULV and thermal fog mosquito control applications in temperate and desert environments. *Journal of the American Mosquito Control Association* **26**: 183-97.

Brown, J. R., Mickle, R. E., Yates, M. and Zhai, J. (2003) Optimizing an aerial spray for mosquito control. *Journal of the American Mosquito Control Association* **19**: 243-50.

Brown, J. R., Williams, D. C., Gwinn, T. and Melson, R. O. (1998) Flit-gun sprayer characteristics. *Rev Panam Salud Publica/Pan American Journal of Public Health* **3**: 322-5.

Brown, J. R., Rutledge, C., Reynolds, W. and Dame, D. A. (2006) Impact of low aerial application rates of Dibrom 14 on potential vectors. *Journal of the American Mosquito Control Association* **22**: 87-92.

Chadwick, P. R. (1975) The activity of some pyrethroids, DDT and lindane in smoke from coils for biting inhibition, knock-down and kill of mosquitoes (Diptera. Culicidae). *Bulletin of Entomological Research* **65**: 97-107.

Cilek, J. E. (2008) Application of insecticides to vegetation as barriers against host-seeking mosquitoes. *Journal of the American Mosquito Control Association* **24**: 172-6.

Clayson, P. J., Latham, M., Bonds, J. A. S., Healy, S. P., Crans, S. C. and Farajollahi, A. (2010) A droplet collection device and support system for ultra-low-volume adulticide trials. *Journal of the American Mosquito Control Association* **26**: 229-32.

Curtis, G. A. and Beidler, E. J. (1996) Influence of ground ULV droplet spectra on adulticide efficacy for *Aedes taeniorhychus*. *Journal of the American Mosquito Control Association* **12**: 368-71.

Dukes, J., Zhong, H., Greer, M., Hester, P. and Hogan, D. (2001). Downwind movement and deposit of insecticides applied by fixed wing aircraft for adult mosquito control. *Wing Beats* **12**: 18-25.

Dukes J., Zhong, H., Greer, M., Hester, P., Hogan, D. and Barber, J. A. S. (2004) A comparison of two ultra-low-volume spray nozzle systems by using a multiple swath scenario for the aerial application of fenthion against adult mosquitoes. *Journal of the American Mosquito Control Association* **20**: 36-44.

Elnaiem, D. A., Kelley, K., Wright, S., et al. (2008) Impact of aerial spraying of Pyrethrin insecticide on *Culex pipiens* and *Culex tarsalis* (Diptera: Culicidae) abundance and West Nile Virus infection rates in an urban/suburban area of Sacramento County, California. *Journal of Medical Entomology* **45**: 751-7.

Emerson, P. M., Lindsay, S. W., Walraven, G. E., et al. (1999) Effect of fly control on trachoma and diarrhoea. *Lancet* **24**: 1401-3.

Esu, E., Lenhart, A., Smith, L. and Horstick, O. (2011) Effectiveness of peridomestic space spraying with insecticide on dengue transmission: systematic review. *Tropical Medicine and International Health* **15**: 619-31.

Gerry, A. C., Zhang, X., Leng, G., Inman, A. D. and Krieger, R. I. (2005) Low pilot exposure to pyrethrin during ultra-low-volume (ulv) aerial insecticide application for control of adult mosquitoes. *Journal of the American Mosquito Control Association* **21**: 291-5.

Gratz, N. G. (1999) Space sprays for the control of *Aedes aegypti* in South-East Asia and the Western Pacific. *Dengue Bulletin* **23**: 80-4.

Hadaway, A. B. and Barlow, F. (1965) Studies on the deposition of oil drops. *Annals of Applied Biology* **55**: 267-74.

Haile, D. G., Mount, G. A. and Pierce, N. W. (1982) Effect of droplet size of malathion aerosols on kill of caged adult mosquitoes. *Mosquito News* **42**: 576-83.

Hoffmann, W. C., Walker, T. W., Smith, V.L., Martin, D. E. and Fritz, B. K. (2007a) Droplet size characterization of handheld atomization equipment typically used in vector control. *Journal of the American Mosquito Control Association* **23**: 315-20.

Hoffmann, W. C., Walker, T. W., Martin, D. E., et al. (2007b) Characterization of truck-mounted atomization equipment typically used in vector control. *Journal of the American Mosquito Control Association* **23**: 321-9.

Hoffman, W. C., Fritz, B. K., Farooq, M. and Cooperband, M. F. (2008a) Effects of wind speed on aerosol spray penetration in adult mosquito boassay cages. *Journal of the American Mosquito Control Association* **24**: 419-26.

Hoffmann, W. C., Walker, T. W., Fritz, B. K., et al. (2008b) Spray characterization of thermal fogging equipment typically used in vector control. *Journal of the American Mosquito Control Association* **24**: 550-9.

Hoffmann, W. C., Walker, T. W., Fritz, B. K., et al. (2009a) Spray characterization of ultra-low-volume sprayers typically used in vector control. *Journal of the American Mosquito Control Association* **25**: 332-7.

Hoffmann, W. C., Farooq, M., Walker, T. W., et al. (2009b) Canopy penetration and deposition of barrier sprays from electrostatic and conventional sprayers. *Journal of the American Mosquito Control Association* **25**: 323-31.

Hornby, J. A., Robinson, J., Opp, W. and Sterling, M. (2006) Laser-diffraction characterization of flat-fan nozzles used to develop aerosol clouds of aerially applied mosquito adulticides. *Journal of the American Mosquito Control Association* **22**: 702-6.

Johnstone, D. R., Cooper, J. F., Casci, F. and Dobson, H. M. (1990) The interpretation of spray monitoring data in tsetse control operations using insecticidal aerosols applied from aircraft. *Atmospheric Environment* **24A**: 53-61.

Koenraadt, C. J. M., Aldstadt, J., Kuchalao, U., Kengluecha, A., Jones, J. W. and Scott, T. W. (2007) Spatial and temporal patterns in the recovery of *Aedes aegypti* (Diptera: Culicidae) populations after insecticide treatment. *Journal of Medical Entomology* **44(1)**: 65-71.

Latta, R., Anderson, A. D., Rogers, E. E., et al. (1947) The effect of particle size and velocity of movement of DDT aerosols in a wind tunnel on the mortality of mosquitoes. *Journal of the Washington Academy of Science* **37**: 397-407.

Lee, C. W., Pope, G. G., Kendrick, J. A., Bowles, G. and Wiggett, G. (1975) Aerosol studies using an Aztec aircraft fitted with Micronair Equipment for tsetse fly control in Botswana. *Centre for Overseas Pest Research Miscellaneous Report No. 118.*

Lofgren, C. S., Ford, H. R., Tonn, R. J. and Jatanasen, S. (1970) The effectiveness of ultra-low volume applications of malathion at a rate of 6 US fluid ounces per acre in controlling *Aedes aegypti* in a large-scale test at Nakorn Sawan, Thailand. *Bulletin of the World Health Organisation* **42**: 15-25.

Lothrop, H. D., Huang, H. Z., Lothrop, B. B., Gee, S., Gomsi, D. E. and Reisen, W. K. (2007) Deposition of pyrethrins and piperonyl butoxide following aerial ultra-low volume applications in the Coachella valley, California. *Journal of the American Mosquito Control Association* **23**: 213-19.

Lothrop, H., Lothrop, B., Palmer, M., et al. (2007) Evaluation of pyrethrin aerial ultra-low volume applications for adult *Culex tarsalis* control in the desert environments of the Coachella valley, Riverside county, California. *Journal of the American Mosquito Control Association* **23**: 405-19.

Macedo, P. A., Schleier III, J. J., Reed, M., et al. (2010) Evaluation of efficacy and human health risk of aerial ultra-low volume application of pyrethrins and piperonyl butoxide for adult mosquito management in response to West Nile virus activity in Sacramento County, California. *Journal of the American Mosquito Control Association* **26**: 57–66.

Matthews, G. A., Dobson, H. M., Nkot, P. B., Wiles, T. L. and Birchmore, M. (2009) Preliminary examination of integrated vector management in a tropical rain forest area of Cameroon. *Transactions of the Royal Society of Tropical Medicine and Hygiene* **103**: 1098-104.

Moore, J. C., Dukes, J. C., Clark, J. R., Malone, J., Hallmon, C. F. and Hester, P. G. (1993) Downwind drift and deposition of malathion on human targets from ground ultra-low volume mosquito sprays. *Journal of the American Mosquito Control Association* **23**: 190-207.

Mount, G. A. (1998) A critical review of ultra-low-volume aerosols of insecticide applied with vehicle-mounted generators for adult mosquito control. *Journal of the American Mosquito Control Association* **14**: 305-34.

Mount, G. A., Biery, T. L. and Haile, D. G. (1996) A review of ultra-low-volume aerial sprays of insecticide for mosquito control. *Journal of the American Mosquito Control Association* **12**: 601-18.

Mount G. A., Lofgren, C. S., Pierce, N. W. and Husman, C. N. (1968) Ultra-low-volume non-thermal aerosols of malathion and naled for adult mosquito control. *Mosquito News* **28**: 99-103.

Pant, C. P., Mount, G. A., Jatanasen, S. and Mathis, H. L. (1971) Ultra-low-volume ground aerosols of technical malathion for the control of *Aedes aegypti* L. *Bulletin of the World Health Organisation* **45**: 805-17.

Perich, M. J., Rochan, O., Castro, J. L., et al. (2003) Evaluation of the efficacy of lambda cyhalothrin applied by three spray application methods for emergency control of *Aedes aegypti* in Costa Rica. *Journal of the American Mosquito Control Association* **19**: 58-62.

Peterson, R. K. D., Macedo, P. A. and Davis, R. S. (2006) A human-health risk assessment for West Nile virus and insecticides used in mosquito management. *Environmental Health Perspectives* **114**: 366-72.

Rathburn, C. B. and Dukes, J. C. (1989) A comparison of the mortality of caged adult mosquitoes to the size, number and volume of ULV spray droplets sampled in an open and a vegetated area. *Journal of the American Mosquito Control Association* **5**: 173-5.

Reiter, P. (1983) A portable, battery-powered trap for collecting gravid *Culex* mosquitoes. *Journal of the American Mosquito Control Association* **43**: 496-8.

Robinson, J. (2001) Light at the end of a wind tunnel. *Wing Beats* **12**: 10-17.

Schleier III, J. J. and Peterson, R. K. D. (2010) Deposition and air concentrations of permethrin and naled used for adult mosquito management. *Archives of Environmental Contamination and Toxicology* **58**: 105-11.

Scott, T. W., Morrison, A. C., Lorenz, L. H., et al. (2000) Longitudinal studies of *Aedes aegypti* (Diptera. Culicidae) in Thailand and Puerto Rico: population dynamics. *Journal of Medical Entomology* **37**: 77-88.

Service, M. W. (1993) *Mosquito Ecology: Field Sampling Methods.* Chapman & Hall, London.

Simpson, J. E. (2006) Emergency mosquito aerial spray response to the 2004 Florida Hurricanes Charley, Frances, Ivan, and Jeanne: An overview of control results. *Journal of the American Mosquito Control Association* **22**: 1457-63.

Sudia W. D. and Chamberlain R. W. (1962) Battery operated light trap, an improved model. *Mosquito News* **22**: 126-9.

Trout, R. T., Brown, G. C., Potter, M. F. and Hubbard J. L. (2007) Efficacy of two pyrethroid insecticides applied as barrier treatments for managing mosquito

(Diptera: Culicidae) populations in suburban residential properties. *Journal of Medical Entomology* **44**: 470–7.

Vazquez-Prokopek, G. M., Galvin, W. A., Kelly, R. and Kitron, W. (2009) A new cost-effective, battery-powered aspirator for adult mosquito collection. *Journal of Medical Entomology* **46**: 1256–7.

Weissmann, G. (2008) Gore's Fever and Dante's Inferno: Chikungunya Reaches Ravenna. *The Journal of the Federation of American Societies of Experimental Biology* **22**: 635–8.

White, S. L., Ward, M. P., Budke, C. M., Cyr, T. and Bueno, R. (2009) A comparison of gravid and under-house CO_2-baited CDC light traps for mosquito species of public health importance in Houston, Texas. *Journal of Medical Entomology* **46**: 1494–7.

WHO (2009) 'Guidelines for efficacy testing of insecticides for indoor and outdoor ground-applied space spray applications.' WHO/HTM/NTD/WHOPES/2009.2

WHO (2010) 'Generic risk assessment model for indoor and outdoor space spraying of insecticides.' WHO/HTM/NTD/WHOPES/2010.6 (web only).

Yeomans, A. H., Rogers, E. and Ball, W. H. (1949) Behaviour of aerosol droplets. *Journal of Economic Entomology* **42**: 591–6.

Zhong, H., Latham, M., Payne, S. and Brock, C. (2004) Minimizing the impact of the mosquito adulticide naled on honey bees, *Apis mellifera* (Hymenoptera: Apidae): aerial ultra-low-volume application using a high-pressure nozzle system. *Journal of Economic Entomology* **97**: 1–7.

Chapter 4

Bed Nets and Treated Clothing

Use of bed nets was observed by the Greek historian Herodotus (circa 484-425 BCE) in Egypt. Nets made from silk were used in Ancient China (Figure 4.1), while nets were also made with fibres from Ramie – *Boehmeria nivea,* known as 'China grass'. The use of curtains around beds dates back many centuries, with four-poster beds used by the more affluent of society to keep out the cold, but may also have helped keep out anopheline mosquitoes. Explorers and others working in the tropics always took bed nets as an essential part of their kit. Ronald Ross, who was subsequently knighted and awarded a Nobel Prize, suffered malaria in India despite sleeping under a net, while he was researching the connection between mosquitoes and malaria. In 1897, he made his landmark discovery, when he found the malaria parasite in stomach tissue while dissecting an anopheline mosquito fed four days earlier on a person suffering from malaria. In the absence of modern insecticides, mosquito control was then mainly by drainage to reduce breeding and protecting people by constructing houses with mosquito mesh over windows and around verandahs, plus the use of bed nets (Figure 4.2).

While untreated bed nets have been used for many centuries, their treatment with insecticide was new (Lindsay and Gibson, 1988; Carnevale et al., 1988), although earlier Buxton (1945) referred to another promising use of DDT 'by impregnating of relatively wide meshed bed nets which would make them effective barriers against small species of *Anopheles*.' When WHO stopped promoting indoor residual spraying (IRS), the number of cases of malaria increased, so although very few people slept under bed nets in Africa, research began on treating nets with insecticide. Initially the studies involved treating individual nets (insecticide treated net – ITN) by soaking them with an insecticide formulation diluted in water. Generally these treated nets have to be re-treated every 6 months, or after about three washes, but this is rarely done in practice, although kits were introduced in an attempt to get people to re-impregnate their nets. However, new technology has led to the development

Integrated Vector Management: Controlling Vectors of Malaria and Other Insect Vector Borne Diseases, First Edition. Graham Matthews.
© 2011 John Wiley & Sons, Ltd. Published 2011 by John Wiley & Sons, Ltd.

(a)

(b)

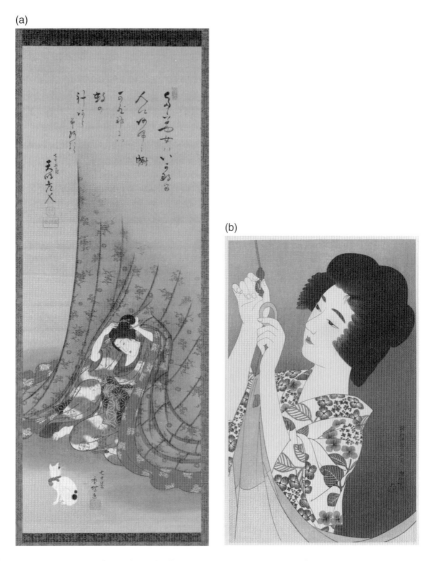

Figure 4.1   (a) Mosquito net as depicted in eleventh-century literary work, *The Tale of Genji* from a Japanese Scroll of the Edo period 1855 (Freer Gallery of Art, Smithsonian Institute, Washington, DC: purchase – Harold P. Stern Memorial Fund, F1995.–17, reproduced with permission); (b) *Lady hanging the net* (Freer Gallery of Art, Smithsonian Institute, Washington, DC: purchase – Harold P. Stern Memorial Fund, S2003.8.303, reproduced with permission).

of bed nets that self-regenerate after washing, referred to as long-lasting insecticidal nets (LNs, sometimes LLINs). Two types of technology exist, whereby the insecticide is impregnated onto polyester fibres using a wash resistant resin, or incorporated directly inside PE or polypropylene yarns.

A person sleeping under a bed net acts as a strong attractant to lure mosquitoes to rest on the net, where they can pick up a lethal dose of insecticide. Even if a treated net gets torn, mosquitoes are still likely to get a

Figure 4.2 Different styles of bed net (reproduced from Chavasse et al., 1999, with permission of WHO).

lethal dose before finding an opening in the net (Lines et al., 1987). The net is also an effective barrier preventing bites, unless the person has uncovered skin touching the net. Their use is justified as most biting occurs between 2200 and 0500 hours, although many people are exposed earlier in the evening or when they get up early in the morning (Maxwell et al., 1998). People are also exposed to mosquito bites when they visit latrines during the night, as these are invariably outside the houses in rural areas. In Brazil, where *An. darlingi* is a key vector, it has been suggested that use of insecticide treated bed nets (ITN) may not be as effective as other control strategies where the vector can bite earlier than 2200 hours and a high outdoor-to-indoor biting ratio has been noted (Ferreira and da Silva-Nunes, 2010; Tadei et al., 1998).

Material

At present it is possible to obtain nets already impregnated with insecticide, but initially nets had to be treated by soaking them in a container with a pyrethroid insecticide mixed with water and then allowing them to dry away from direct sunlight. A guide for treating nets was published by WHO (2002a). The persistence of the deposit on the nets was generally expected to be at least 6 months, sufficiently long for protection during a single rainfall season. Tablets or sachets of the insecticide were commercially available for re-treatment of these nets. Latterly, products were specially formulated to bind to the net and persist when the net was washed. Yates et al. (2005) reported that by using a tablet of deltamethrin (KO-Tab 1-2-3), the insecticide content of a treated net was halved within 20 washes, but there was no loss of biological efficacy in 3-min exposure bioassays, even after 30 washes. However, this technique does not convert a net into a long-lasting net and concerns about exposing individuals to the insecticide during treatment of individual nets remained, so better alternatives were needed. Nevertheless, individuals can treat their own net.

Subsequently, a new type of net was developed, in which the insecticide was impregnated into the fibre before the netting was made and cut up to make the bed nets. The technique is to mix the insecticide with a resin or binder prior to extrusion of the fibre with an appropriate thickness that is then woven into the fabric. The success depends on the dynamic behaviour of the insecticide in the fibre, so that insects touching the fibre surface contact the insecticide. The release rate, especially after washing of the nets, is an important factor to maintain the effect over a prolonged period. In some cases the insecticide is mixed with the polymer before the fibres are extruded, while other nets have the material treated with a surface coating of resin and insecticide:

- *Coating technology*: the fibre is extruded and the insecticide is then contained in a resin that is coated to the exterior of the fibres. The insecticide must migrate through the thin resin coating to the fibre surface to be bio-available.
- *Incorporation technology*: the insecticide (and synergist) are added to the mixture prior to extrusion and thus are distributed throughout the polymer matrix. The insecticide must migrate from throughout the polymer to the fibre surface to be bio-available.

Instead of the traditional cotton nets, most mosquito nets are now made with a polyester (PET), polyethylene (PE) or polypropylene lightweight fibre (Curtis et al., 1996), as cotton fibre is unsuitable in making LLINs. These new nets are available in different mesh sizes and thread thickness. These synthetic fibres are generally more durable than cotton and less likely to tear, but much depends on how they are used and how they are manufactured. Different manufacturers have their own specification in terms of mesh size and denier, which affects the overall integrity of a net (denier is the weight in grams of 9,000 m of yarn; 9,000 m of a single strand of silk weighs 1 g). Nets also vary, because they can be made with a monofilament (PE) or have a multifilament (PET) construction due to the technology used in constructing the nets with different polymers. After washing, a surface film is regenerated from the insecticide within the polymer filament. Currently, WHO requires that a LLIN should last a minimum of 3 years. According to Skovmand and Bosselmann (2011) measurements of tensile strength using a hook and clamp indicated that commercial monofilament yarn polyethylene nets are significantly stronger than multi-filament polyester nets.

Dabire et al. (2006) compared bed nets made from PT and PE in a villages trial that showed no difference between the two types of net over a 2-year study. In addition to mortality of mosquitoes, fewer were detected in houses compared with controls, indicating a deterrent effect.

Lindblade et al. (2005) defined failure when there was less than 50% mortality in the first of two consecutive monthly bioassays using cone tests on several different types of nets distributed in Western Kenya. Using the criteria, between 8 and 50% of the nets failed over the test period, during which mortality varied, partly due to the effect of washing.

Mesh size

A key complaint of sleeping under a bed net is that ventilation is restricted, which can result in user rejection, so the net needs to have a mesh size that is large enough for air circulation but as small as possible to keep out mosquitoes. An opening of 1.2 mm × 1.2 mm with 24 holes/cm² has been recommended. Smaller holes (0.6 mm × 0.6 mm) are needed to protect against sand flies. Zollner et al. (2007) noted that some of the smaller species of sand fly will pass through all but the finest mesh, such as the standard US military mesh, which has 113 holes/cm². Some nets have slightly larger mesh size (up to

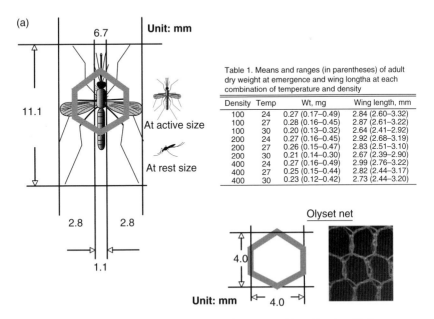

Table 1. Means and ranges (in parentheses) of adult dry weight at emergence and wing longtha at each combination of temperature and density

Density	Temp	Wt, mg	Wing length, mm
100	24	0.27 (0.17–0.49)	2.84 (2.60–3.32)
100	27	0.28 (0.16–0.45)	2.87 (2.61–3.22)
100	30	0.20 (0.13–0.32)	2.64 (2.41–2.92)
200	24	0.27 (0.16–0.45)	2.92 (2.68–3.19)
200	27	0.26 (0.15–0.47)	2.83 (2.51–3.10)
200	30	0.21 (0.14–0.30)	2.67 (2.39–2.90)
400	24	0.27 (0.16–0.49)	2.99 (2.76–3.22)
400	27	0.25 (0.15–0.44)	2.82 (2.44–3.17)
400	30	0.23 (0.12–0.42)	2.73 (2.44–3.20)

Figure 4.3 (a) Diagram showing a mosquito on an 'Olyset' net, with larger mesh openings than other types of bed net. Note, the anopheline mosquito wing length can vary from 2.39–3.32 mm; (b) making an Olyset net in Tanzania; (c) Olyset net in use; (d) Olyset net in hospital (photos: Sumitomo).

(b)

(c)

(d)

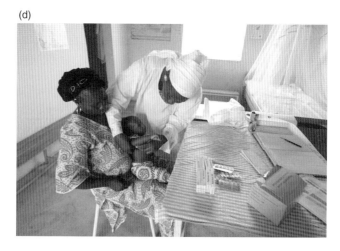

Figure 4.3 (*Continued*)

Table 4.1 Mesh size of some commercially available nets.

Net	Holes/cm²
Olyset	6.25
Dura net	min. 20
NetProtect	20 or 30
Interceptor	24
PermaNet	25

4 mm × 4 mm) for better ventilation, but any mosquito attempting to get through this mesh will inevitably touch the treated fabric (Figure 4.3a). This will depend especially on how fast the regeneration of insecticide on the surface of the nets occurs after washing. The number of holes per cm² for some commercial nets is shown in Table 4.1.

Shape

The net needs to be hung over a bed, so that the occupant is not touching the bed-net fabric while sleeping, otherwise mosquitoes will find exposed skin surfaces and blood feed. Similarly, the bottom of the net has to be tucked under the mattress to avoid mosquitoes entering the net. Rectangular mosquito nets (Figure 4.2) offer the roomiest protection around a sleeping person, whereas the slanted sides of tent-shaped or pyramidal nets allow for less space but are easier to hang as they require only a single point of attachment to the roof or ceiling. They are also more suitable in round houses. However, conical nets are more complicated and expensive to make. To hang the net correctly, hooks or a nail may be needed and string to tie the corners firmly above the bed.

Insecticide

The choice of insecticide for treating bed nets has so far been limited to pyrethroids. Nets treated with α-cypermethrin, deltamethrin, λ-cyhalothrin and permethrin (Table 4.2) have been marketed, but the great concern is the risk of resistance to pyrethroids, particularly as these insecticides are also used in agriculture and for IRS (Chapter 2). Experimental tests have been made with some other insecticides, notably carbosulfan, pirimiphos methyl and chlorpyrifos-methyl and chlorfenapyr. None of the formulations used were wash resistant, so whether these may be used in the future as they are more toxic is debatable, although in theory they would be useful where resistance has already been detected. However, pyrethroids have been the primary insecticide used because of their low toxicity, as using other groups of insecticides on nets may create toxicity problems.

Before the factory-produced ITNs (Figure. 4.3b) were available, the recommendation was to soak a net in an insecticide and then dry the net in the shade (Figure 4.4e, f).

Curtis et al. (1996) reported that nets dipped with permethrin with cis:trans isomer ratios 25:75 and 40:60 were equally effective, and that a lower rate of permethrin (200 mg/m²) was as effective as 500 mg/m². However, dipped nets do not provide long-lasting activity, as the insecticide residue is readily removed by washing, and typically such nets need re-treatment after 6 months. In these bioassay tests, λ-cyhalothrin deposits showed prolonged activity, except on PE netting. λ-cyhalothrin CS at the very low rate of 3 mg/m² gave good results, including after washing and re-treatment. Holed nets treated with either cyfluthrin

Table 4.2   Insecticides currently recommended by WHO for treating bed nets.

Insecticide	Formulation	Dosage g/m² of netting
α-cypermethrin	SC 10%	20–40
Cyfluthrin	EW 5%	50
Deltamethrin	SC 1%; WT 25%	15–25
	WT 25% + binders	
Etofenprox	EW 10%	200
λ-cyhalothrin	CS 2.5%	10–15
Permethrin	EC 10%	200–500

EC = emulsifiable concentrate; EW = emulsion, oil in water; CS = capsule suspension; SC= suspension concentrate; WT = water dispersible tablet.
Lambda-cyhalothrin 10% CS + binder (ICON_ MAXX) Target dose of 50 mg/m².

(a)

Figure 4.4   (a) Pallets of 'Permanet' bed nets ready for distribution. These were donated by the charity (Against Malaria Foundation) and distributed in Cameroon by the Yaounde Initiative Foundation; (b) training villagers on how to put a net over a bed; (c) distributing nets to a villager (photos: Didier Baleguel); (d) bed net in use in a Cameroonian house; (e) treating a net with insecticide in a village; (f) hanging the treated net to dry in shaded area.

(b)

(c)

(d)

Figure 4.4 (*Continued*)

(e)

(f)

Figure 4.4 (*Continued*)

(5 EW formulation applied at the rate of 50 mg a.i./m²) or λ-cyhalothrin (2.5 CS formulation at 10 mg a.i./m²) performed well after 15 months of domestic use.

Asidi et al. (2004) reported tests in which carbosulfan applied at 200 mg/m² was significantly better than pyrethroid treated nets against *Anopheles gambiae* in Cote 'Ivoire. However, the deposit on the net was less persistent than a pyrethroid treated net. Also, it is considered that carbosulfan is unlikely to be used commercially because of its toxicity.

Due to resistance to pyrethroids, nets impregnated with λ-cyhalothrin killed only 30% of *An. gambiae* and 8% of *Culex quinquefasciatus*, while nets impregnated with chlorpyrifos methyl, also applied using a micro-encapsulated formulation, killed more (45% of *An. gambiae* and 15% of *Cx. quinquefasciatus*), but it was not very persistent (N'Guessan et al., 2010a).

The synergist piperonyl butoxide (PBO) has been used extensively with pyrethroid insecticides applied as sprays to overcome the resistance problem to some extent, as PBO is an inhibitor of mixed function oxidases implicated in pyrethroid resistance. A new net now incorporating a synergist PBO is mixed with deltamethrin and used on the top panel of a net (Tungu et al., 2010), but in initial trials the synergism on unwashed nets was lower than expected, possibly due to an unidentified resistance mechanism that was not affected by PBO (N'Guessan et al., 2010b).

Insecticide impregnation

The availability of nets impregnated with insecticide, referred to as long-lasting insecticide treated nets (LNs or LLINs), has given a major advantage over the nets that require regular re-treatment. To market these nets, manufacturers have to establish that they are effective over several years, according to established WHO guidelines, to determine if the nets meet the criteria and therefore the definition of an LLIN. Essentially, WHO arranges for laboratory tests, including a sequence of washings, and small-scale field trials in order to give interim recommendations, which are required by donors to buy the nets. A standard washing method has been developed by WHO for conducting wash resistance tests in the laboratory. However, Atelier et al. (2010a) observed that nets washed by the standard WHO protocol, were more effective in killing mosquitoes than nets washed by local methods of hand rubbing and beating the nets on rocks.

Small-scale trials are then followed by community-scale trials (Phase 3) for a minimum of 3 years to determine if the product meets the current WHO definition of an LLIN. WHO requires tests on these nets with a sequence of washings to determine the life expectancy of the treated net. Reports on individual types of net submitted to the WHO Pesticide Evaluation Scheme are published.

In February 2010, the list of long-lasting ITNs that had been evaluated and were approved, either as full or interim WHO recommendations, was as follows:

- *DawaPlus*® *2.0* Deltamethrin coated on PT Interim
- *Duranet*® α-cypermethrin incorporated into PE Interim
- *Interceptor*® α-cypermethrin coated on PET Interim
- *Netprotect*® Deltamethrin incorporated into PE Interim
- *Olyset*® Permethrin incorporated into PE Full
- *PermaNet*® *2.0* Deltamethrin coated on PET Full
- *PermaNet*® *2.5* Deltamethrin coated on PET with strengthened border Interim
- *PermaNet*® *3.0* Combination of deltamethrin coated on PET with strengthened border (side panels) and deltamethrin and PBO incorporated into PE (roof) Interim.

The WHO web page for currently recommended ITNs is: http://www.who.int/whopes/Long_lasting_insecticidal_nets_Aug09.pdf

WHO has published a generic guide for risk assessment when treating bed nets with insecticides and their subsequent use.

Impact of washing nets

There have been reports that an LLIN is not so effective immediately after a washing, but the insecticide within the fibre re-establishes a toxic dose on the surface of the netting very quickly. Each net has a different regeneration time. In general, PET nets regenerate faster than PE nets A laboratory method has been developed for WHO using a Julabo® SW22 water bath shaker (made in Germany), in which pieces of net are shaken during the test. Kayedi et al. (2009) has reported on a series of tests in which three pieces of netting (30 cm × 35 cm) were cut from a rectangular net with one piece from the top, one from one of the long sides and one from one of the short sides. Each sample of netting was placed in a 1-litre glass bottle to which 500 ml of deionised water with 2 g/l of soap 'savon de Marseille' (pH 10.2) was added. The bottles were placed on the water bath shaker and agitated with 155 strokes per min at 30°C for 10 min. After washing with soap, the pieces of net were rinsed twice with deionised water at 30°C for 10 min each, at the same agitation speed, before being dried on a line in a room at 27°C and 80% relative humidity (RH), with artificial light (night/day 12 h/12 h). These pieces of nets were washed up to 21 times, with a 2-day interval between washes. The efficacy of the nets was assessed using two types of bioassay (mean median knock-down times and mortality 24 h after a 3-min exposure) with reared female *Anopheles stephensi*. Although there was no intense rubbing of the nets, insecticide was removed from the net at each wash and efficacy declined, which may not be truly representative of what happens in the field. However, tests are repeatable. The deltamethrin treated PermaNet 2.0 was the most effective and this corresponded with similar tests (Gimnig et al., 2005), in which efficacy was tested weekly with *An. gambiae* (Kisumu strain) using WHO cone bioassays. The PermaNet 1.0 was the most wash resistant, with more than 50% mosquito mortality in WHO cone bioassays after as many as 20 washes. Most nets lost more than 90% of their biological activity after 6 washes, as measured by 24-h mortality of *An. gambiae* in WHO cone tests and all nets lost more than 50% of their initial insecticide concentrations after 20 washes, except for the Olyset® net.

Tests have compared PET nets with PE and cotton nets treated with K-O Tab 1-2-3 and washed 20 times (Oxborough et al., 2009). The performance of PE matched that of PET in all bioassays, whereas low mortality and knock-down occurred with cotton and nylon nets in cone and cylinder bioassays. After 20 washes, PET nets had retained 16.5% of the loading dose of deltamethrin compared with 28.7% on PE, 38.9% on cotton and 2.2% on nylon. Although the cotton nets retained a high concentration of insecticide, it seems that the insecticide was bound within the cotton fibres rather than on the surface.

Oxborough et al. (2009) also reported results with a tunnel test (WHO, 2005), in which samples of the nets (1cm diameter) were fitted in frames across the tunnels. Unfed female pyrethroid susceptible *An. gambiae* (Kisumu strain) mosquitoes were left in the tunnels for 12h (26 ± 2°C and relative humidity of 70–80%) overnight. Next day mortality was recorded and any live mosquitoes were transferred to plastic cups with sugar solution to assess delayed mortality. These tests showed ≥85% mortality following 20 washes for KO Tab 1-2-3 treated PET, PE and cotton, but only 63% mortality on nylon.

These tests indicate that nets, even labelled 'long lasting', will decrease in their effectiveness, depending on the number of times they are washed and the extent to which they are rubbed during washing. Atieli et al. (2010a) pointed out that the standard WHO Pesticide Evaluation Programme (WHOPES) washing protocol under-estimates the amount of insecticide washed from LLINs compared to the abrasive washing procedures that can be used in the field and that the nets should be dried in the shade. Thus, advice on washing should be given during distribution of the nets. In experiments reported by Atieli et al. (2010b), on differences between LLINs from different manufacturers, the nets were washed twice weekly, which is unrealistic as the insecticide within the fibre has insufficient time to regenerate a stable deposit on the net surface. Nevertheless, if nets are washed too frequently, their effectiveness will decline quicker, as the regenerated insecticidal deposit is washed from the surface of the net. Thus, all nets will require re-treatment if they are otherwise not torn and can be re-used. However, PE nets are more difficult to re-treat than PET nets, due to the type of binder required to stick the insecticide to the fibre.

Bed nets can be evaluated using cone tests, tunnel tests and in experimental huts. The cone test is aimed at assessing the mortality of mosquitoes exposed to a net surface, but if the insecticide is repellent, it is not really suitable. The surface dose is high with new nets and will be low directly after washing, so mortality may be low if the nets are tested too early before regeneration occurs.

As permethrin is highly repellent, mosquitoes sometimes do not rest on the treated net and with LLINs such as the Olyset net, the relatively low surface dose does not kill very effectively. This is evident in tunnel tests with washed nets. These tests are a much better indicator of performance for insecticides in general and particularly permethrin, but they take longer to do. Ideally treated nets need to be evaluated in experimental huts, followed by cross-sectional epidemiological trials in areas where nets are used in whole villages.

Distribution of nets

Initially the aim was to protect young children (<5 years), who are most likely to die as they had not built up any immunity to infection (Alonso et al., 1991). It has also been considered important to protect pregnant women. Various charities and donors have promoted the use of nets to reduce child mortality. Major efforts by WHO with the Roll-Back Malaria (RBM) programme and the Global Fund for AIDs, TB and malaria both having a significant impact in encouraging people to sleep under ITNs, especially as the impact on the local mosquito population is much greater when there are more nets in a locality, often referred to as the 'community effect'. Nevertheless, a large proportion of the population in some countries does not have a net or does not wish to sleep under a net.

There have been various ways in which bed nets have been distributed (Figures 4.4a, b, c). The cost of an individual net, even without insecticide impregnation, is high for the majority of the population, so considerable lobbying of governments and charities by advocacy groups has argued that nets should be distributed free, especially for use by children. Curtis et al. (2003) argued that supplying ITNs free and replacing them after about 4 years provided a more comprehensive and equitable coverage of the population than had been reported for social marketing systems. In a study in Tanzania, only 9.3% of people used nets where they had to be purchased, in contrast to over 90% in areas where they had been issued free (Maxwell et al., 2006). Pettifor et al. (2009) confirmed that in the Democratic Republic of Congo, freely distributed nets result in high usage and that distribution at antenatal clinics was a highly effective means of distribution. In some countries, distribution of nets has been linked with immunisation programmes, so that when mothers bring a child in for immunisation, an ITN is provided at the same time. When a country aims at universal coverage with bed nets, it has been suggested that there should be a quantification factor of 1.78 people per net, plus a further allowance to compensate for problems during distribution (Kilian et al., 2010).

Unfortunately, in some areas, when subsequent surveys were carried out to determine the extent of their actual use, only a small percentage of the nets were being used. This was often because there had been insufficient instructions given and the recipients lacked the means of hanging the net correctly over a child's bed. In southwest Ethiopia, only 25% of over 4,000 households in a survey of 22 *Gats* (villages) had a functional long-lasting insecticide treated bed net and of these only about 30% were hung correctly (Deribew et al., 2010). There was no significant reduction in the prevalence of malaria among children, as adults were mostly likely to use the nets.

The actual cost of treated nets depends on the type of net, quantities involved and their distribution, the latter being about 40% of the total cost. Large quantities of nets are bulky and require large vehicles for transportation, so need to be distributed during the dry season, as roads may become impassable in the wet season. In one area of Kenya, Guyatt et al. (2002) indicated that IRS would appear to be a more cost-effective method of vector control, at least in the area studied.

In another survey, Eng et al. (2010) showed that the percentage of children of less than 5 years of age sleeping under an ITN ranged from 51.5% in Kenya to 81.1% in Madagascar. Children living in households without an ITN ranged from 9.4 to 30.0%, while in households with a net but not hung in the house, it ranged from 5.1% to 16.1%, despite the efforts of the integrated child health campaigns. The percentage of children living in households where an ITN was suspended, but who were not sleeping under it, ranged from 4.3 to 16.4%.

Clearly a key objective is to ensure that nets are distributed as quickly as possible to end-users. One of the difficulties is transportation within Africa is that routes from the coast go inland, but marketing between African countries has been affected by poor road systems. Manufacturing where nets are needed is the obvious solution and Sumitomo Chemical, producers of an LLIN called Olyset Net, have pioneered this approach. In 2003, they formed a joint venture with A to Z Textile Mills, based in Arusha, Tanzania (Figure 4.3b), who were already making untreated bed nets. With some initial financing provided by the Acumen Fund to allow investment in machinery, the technology for manufacturing nets was transferred on a royalty free basis and manufacturing began – initially on a relatively modest scale but, in 2008, the Joint Venture opened a new, purpose-built facility, which is now producing approximately 29 million nets per annum – more than half of the global Olyset output.

In addition, the factory is employing nearly 6,000 people (mostly women), whose wages are supporting over 25,000 people in the Arusha area. A recent Economic Impact study has highlighted the long-term benefits to the local economy of the Arusha operation. As a result of this success story, Sumitomo has also established smaller-scale net stitching operations, again employing local people, in Malawi, Kenya and Ethiopia, and is currently investigating opportunities for duplicating the Arusha venture in West Africa. Other companies, which had previously manufactured LLINs outside Africa, are now also investigating the benefits of local production.

Trial data

There have been many reports on use of ITNs in different areas of the world. Lengeler (2004) published an extensive review of the impact of treated nets or curtains on mortality, malaria illness, malaria parasitaemia, anaemia and spleen rates within strict criteria, so that 14 cluster randomised and 8 individually randomised controlled trials met the criteria. Trials including only pregnant women were excluded. The conclusions were that ITNs provided 17 and 23% protective efficacy compared to no nets or untreated nets, respectively. Thus, about 5.5 lives per 1,000 children protected with ITNs can be saved each year. By using ITNs, the incidence of uncomplicated malarial episodes in areas with stable malaria was reduced by 50 and 39% compared with untreated or no nets, respectively. Nets also reduced the incidence of malaria in areas of unstable malaria. Overall, the conclusion was that ITNs are highly effective in reducing childhood mortality and morbidity due to malaria.

Sharma et al. (2005) reported on a trial involving three villages in India. One village was supplied with PET 156 – mesh nets treated with deltamethrin (as 1 tablet per net), another had untreated nets, while a third had no nets. The village without nets did receive an IRS treatment towards the end of the post-treatment phase, although catches of mosquitoes continued for several months. Bioassays of the treated nets were reported to achieve 100% mortality over 7 months against *An. fluviatilis*, but declined to 71% against *Anopheles culicifacies*. In Tanzania, the performance of 'Olyset' bed nets and nets treated with α-cypermethrin were evaluated in experimental huts. The nets failed to prevent mosquito biting after 1.5 years in the case of α-cypermethrin and after 7 years with an 'Olyset' net, despite many *An. gambiae* and *Anopheles funestus* being killed (Malima et al., 2008). It was concluded that in hut trials, parts of the sleepers' bodies may touch the sides of the net, giving mosquitoes an opportunity to feed through the netting or through holes that develop during use and that laboratory (tunnel) tests over-estimate the field effectiveness of treatments. In practice, treated bed nets need to be repaired and replaced.

Gu and Novak (2009) have pointed out that the effectiveness of treated bed nets depends on the direct impact on individual mosquitoes, including mortality and repellence, which can vary between vector species. They conclude that the success of scaling up of treated bed net usage may vary, depending on local entomological and epidemiological situations. The inference was that a reduction in risk exposure and incidence of malaria depended on high mortality of the local mosquito vectors.

In South America, treated bed nets were shown to increase the mortality of triatomine bugs and give some protection from Chagas disease (Kroeger et al., 1999). There is also data showing impact of bed nets on sand flies and thus on the incidence of leishmaniasis as well as on other insects, such as cockroaches, bed bugs and flies (Chapter 7).

Human landing catches have been used to monitor human biting rates used in determining entomologic inoculation rates (EIR), but with increasing ethical concerns, light traps (Figure 2.17b) were assessed as an alternative method (Fornadel et al., 2010a). It was shown that a Center for Disease Control and Prevention in the USA (CDC) light trap caught nearly twice as many *Anopheles arabiensis* per night as detected by human landing catches and that the results were unaffected by presence of bed nets treated with deltamethrin. Fornadel et al. (2010b) also reported that *An. arabiensis* was caught biting outdoors immediately after sunset and before sunrise, confirming that the protective effects of ITNs was confined to those who remained under a net throughout the period when biting occurred.

Operational use

A guide to National Programme managers was issued to assist in the distribution of bed nets (WHO, 2002b). Subsequently, an edition regarding use of LLINs was issued (WHO, 2007a). According to the WHO Malaria Report for 2009, the number of nets needed to protect all the people at risk of malaria in high-burden countries globally in 2008 was approximately 336 million, assuming that one net covers two persons. The cumulative number of LLINs delivered by manufacturers was 141 million over the period 2006–2008, representing 42% of the number needed in 2008 and assuming a net will last

3 years. According to the WHO, the weighted average estimate of household ITN ownership was 31%, and ITN use by children less than 5 years old was 24% in all 35 high-burden countries in 2008. Since 2008, the supply and use of bed nets has increased significantly, but not everyone likes having a bed net due to its cost, heat, and discomfort when sleeping under it, or they felt there was no need to sleep under an ITN, when they were bitten before they went to bed (Kudom and Mensah, 2010). Nevertheless, the national of scale-up malaria control programmes in some parts of Africa has achieved a substantial impact, with reductions in child mortality, which are consistent with or even greater than the estimated 20% reduction in all child mortality predicted from the controlled trials of ITNs (Steketee and Campbell, 2010).

The WHO Global Malaria Programme has requested national malaria control programmes and their partners involved in insecticide-treated net interventions to:

- purchase only WHO recommended long-lasting insecticidal nets;
- distribute free or highly subsidised LLINs, either directly or through voucher/coupon schemes;
- achieve full LLIN coverage, including in high-transmission areas, by distributing LLINs through existing public health services;
- develop and implement locally appropriate communication and advocacy strategies to promote effective use of LLINs; and
- implement strategies to sustain high levels of LLIN coverage in parallel with strategies for achieving rapid scale-up (WHO, 2007b).

In Ghana, 255 long-lasting bed nets were collected 38 months after distribution to examine their physical condition and residual insecticide levels. Some continued to look in pristine condition, while others were highly damaged with large holes (Figure 4.5) (Smith et al., 2007). Holes were more common near the bottom edge of the net, with 50% having breaks in the seam, while 64% of the nets were repaired by sewing. Analyses of the insecticide deposits showed that 14.9% of the nets had retained their full insecticidal strength. Similarly in Kenya, out of 1,627 nets in a survey of households, 40% were deemed to be of poor quality because of holes (Githinji et al., 2010). Smith et al. (2007) considered that this data provided a guide to those deciding on the purchase of nets, the frequency of replacement and product development. Unfortunately, protection is significantly decreased as the number of holes increases. Training villagers to sew and repair the nets could extend the effective life of these nets.

According to the Malaria Emergency Technical and Operation Response (MENTOR), bed nets are not always effective in refugee camp settings, so they may not be the best tool for preventing malaria. Alternative techniques to protect refugees against malaria from the time they enter the refugee shelter being evaluated, include insecticide-treated blankets, tents, fencing and plastic sheeting.

The cost of using ITNs is discussed in relation to IRS of houses in Chapter 2, and again in Chapter 6, as a part of integrated vector management.

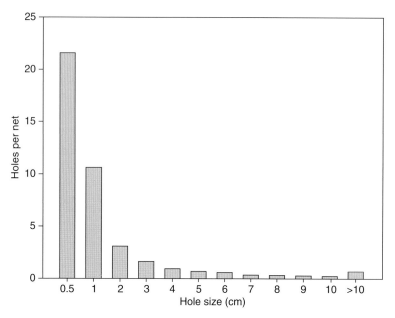

Figure 4.5 Proportion of nets with holes of different size following their use (redrawn from Smith et al., 2007).

Questions have been raised concerning the disposal of treated nets when they are no longer effective. If the net is still without major holes, it could still retain some effectiveness against mosquito bites, but the damaged net could be:

- collected for disposal to a central organisation for incineration, or recycling.
- used as material that could fill holes within a mud-walled house and then coated with mud to restore the wall surface.

Treated clothing

Some insecticides have been impregnated into clothing, especially for military uniforms to protect troops in combat. At present this technique is limited to the use of permethrin, which can be applied to garments by immersion or by coating material before the cloth is cut and made into garments. The latter technique exposes the garment-makers to the insecticide, so while permethrin is of very low mammalian toxicity, appropriate measures to limit occupational exposure may be required. However, this method tends to achieve a more uniform distribution of the insecticide. Re-treatment can be done by spraying the garments (Figure 4.6). Early studies with uniforms impregnated by immersion showed over an 8-week period a significant decrease in soldiers that contracted malaria (14% for soldiers with untreated uniforms to 3%) or similarly leishmaniasis (12% reduced to 3%), with the infected bites being on the unprotected hands or face of those wearing treated uniforms (Soto et al., 1995).

Figure 4.6 Treating military uniforms during operations (reproduced with permission from Army Pest Management Photo Gallery).

Factory-based permethrin coated battle dress uniforms (BDUs), worn during military operations with 50 defined washings, remained effective against *Aedes* mosquitoes, but knock-down activity against *Ixodes* ticks was significantly better with polymer-coated BDUs. Uniforms impregnated by a polymer coating method protected soldiers throughout the effective for the life of the uniform (Faulde et al., 2006).

Battlefield uniforms made from material soaked in permethrin were evaluated in Cote D'Ivoire to assess their field efficacy and their resistance to washing. They provided some protection from mosquito bites but this did not reduce significantly the incidence of malaria. In the same trial, use of the repellent (50% DEET) failed to protect troops after 6 hours, probably due to the high density of *Anopheles* mosquitoes during the night, with repellent action usually less than 3 hours (Deparis et al., 2004).

When comparing the urinary permethrin metabolites from those wearing permethrin treated uniforms for 28 days and a control group over a study period continuing for a further 28 days, the treatment group showed higher metabolite concentrations while the uniforms were worn, but these declined after the 28-day exposure period had ended. Tests showed that the maximum permethrin uptake in the treatment group was below the acceptable daily intake (ADI) (Rossbach et al., 2009), indicating that treated uniforms were safe to use.

In a search for an alternative to the use of permethrin, pirimiphos methyl was evaluated mixed with another repellent KBR3023 (1-(1-methyl-propoxy-carbonyl)-2-(2- hydroxy-ethyl)-piperidine) on uniforms and shown to be more effective than permethrin treated uniforms (Pennetier et al., 2010).

Impregnated sheeting

As an alternative to spraying wall surfaces, it has been suggested that plastic wall and ceiling coverings impregnated with an insecticide would provide a means of avoiding the need to spray inside houses (Diabate et al., 2006). Durable Lining (DL) consists of loosely woven high density PE panels that provide an attractive covering for walls. The panels are made of coloured yarns containing a reservoir of deltamethrin (4.4 g/kg) that provides a long residual life to the lining, as it continually diffuses in a controlled manner to the yarn surface, providing a lethal dose to adult mosquitoes. The insecticide is incorporated into every strand of coloured yarn. Through a controlled migration technology, the insecticide is continually refreshed at the surface of the fabric. Mosquitoes and other insects that come into contact with this treated surface acquire a lethal dose while resting or walking on the lining.

The DLs are supplied with nail caps to allow fixing to the material to walls (Figures 4.7a, b). Manufacturers claim the advantage of DL is that the

(a)

Figure 4.7 Durable linings: (a) Fitting a durable lining to a wall; (b) close-up of nail used to fasten lining to a wall; (c) durable lining covering a window (photos: John Thomas).

(b)

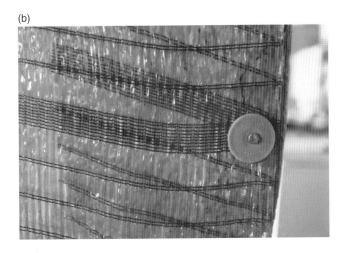

(c)

Figure 4.7 *(Continued)*

insecticide is effective over a longer period than a single IRS application and the factory prepared dosage is more even than wall spraying.

However, apart from the additional cost of the material and difficulty in fitting it in houses of different dimensions, the bulky rolls of material (2.35 m × 100 m) (weight 55 g/m²; 13 kg/roll) may not be easy to transport in parts of Africa. Work is continuing to evaluate the contribution that these wall linings may have in controlling disease vectors. They may have a role when placed in the eaves, as a ceiling and perhaps across windows (Figure 4.7c) or doorways of houses, as they would allow ventilation but kill or prevent entry of mosquitoes into houses.

Conclusion

Bed nets have long been recognised as a means of protection from mosquito bites at night, but the impregnation of the net with insecticide is relatively new. While the technique is very valuable for protection of babies and young children, adults are not always under a net when the vectors of malaria are active. New technology has increased the activity of insecticide on long-lasting nets, but the increased persistence may also increase selection of resistance to the pyrethroid insecticides used to treat nets.

Treatment of uniforms is designed to protect people who are not protected inside buildings and have to be active at night. Durable wall linings have been developed as an alternative to spraying inside houses.

References

Alonso, P., Lindsay, S. W., Armstrong, J. R. M., et al. (1991) The effect of insecticide-treated bed nets on mortality of Gambian children. *Lancet* **337**: 1499-502.

Asidi, A. N., N'Guessan, R., Hutchinson, R. A., Traore-Lamizana, M., Carnevale, P. and Curtis, C. F. (2004) Experimental hut comparisons of nets treated with carbamate or pyrethroid insecticides, washed or unwashed, against pyrethroid-resistant mosquitoes. *Medical and Veterinary Medicine* **18**: 134-40.

Atieli, F. K., Munga, S. O., Ofulla, A. V. and Vulule, J. M. (2010a) Wash durability and optimal drying regimen of four brands of long-lasting insecticide-treated nets after repeated washing under tropical conditions. *Malaria Journal* **9**: 248.

Atieli, F. K., Munga, S. O., Ofulla, A. V. and Vulule, J. M. (2010b) The effect of repeated washing of long-lasting insecticide-treated nets (LLINs) on the feeding success and survival rates of *Anopheles gambiae*. *Malaria Journal* **9**: 304.

Buxton, P. A. (1945) The use of the new insecticide DDT in relation to the problems of tropical medicine. *Transactions of the Royal Society of Tropical Medicine and Hygiene* **38**: 367-93.

Carnevale, P., Robert, V., Boudin, C., et al. (1988) Control of malaria using mosquito nets impregnated with pyrethroids in Burkina Faso. *Bulletin of the Society of Pathol Exot Filiales* **81**: 832-46.

Chavesse, O., Reed, C. and Attawell, K. (1999) *Insecticide-treated net projects: a handbook for managers*. Malaria Consortium 1999.

Curtis, C. F., Myamba, J. and Wilkes, T. J. (1996) Comparison of different insecticides and fabrics for anti-mosquito bed nets and curtains. *Medical and Veterinary Entomology* **10**: 1-11.

Curtis, C. F., Maxwell, C., Lemnge, M., et al. (2003) Scaling-up coverage with insecticide-treated nets against malaria: who should pay? *The Lancet Infectious diseases* **3**: 304-7.

Dabiré, R. K., Diabaté A., Baldet' T., et al. (2006) Personal protection of long-lasting insecticide-treated nets in areas of *Anopheles gambiae* s.s. resistance to pyrethroids. *Malaria Journal* **5**: 12.

Deparis, X., Frere, B., Lamizana, M., et al. (2004) Efficacy of permethrin-treated uniforms in combination with DEET topical repellent for protection of French military troops in Côte d'Ivoire. *Journal of Medical Entomology* **41**: 914-21.

Deribew, A., Alemseged, F., Birhanu, Z., et al. (2010) Effect of training on the use of long-lasting insecticide-treated bed nets on the burden of malaria among vulnerable groups, southwest Ethiopia: baseline results of a cluster randomized trial. *Malaria Journal* **9**: 121.

Diabate, A., Chandre, F., Rowland, M., et al. (2006) The indoor use of plastic sheeting pre-impregnated with insecticide for control of malaria vectors. *Tropical Medicine and International Health* **11**: 597–603.

Eng, J. L. V., Thwing, J., Wolkon, A., et al. (2010) Assessing bed net use and non-use after long-lasting insecticidal net distribution: a simple framework to guide programmatic strategies. *Malaria Journal* **9**: 133.

Faulde, M. K., Uedelhoven, W. M., Malerius, M. and Robbins, R. G. (2006) Factory-based permethrin impregnation of uniforms: residual activity against *Aedes aegypti* and *Ixodes ricinus* in battle dress uniforms worn under field conditions, and cross-contamination during the laundering and storage process. *Military Medicine* **171**: 472–77.

Ferreira, M. U. and da Silva-Nunes, M. (2010) Evidence-based public health and prospects for malaria control in Brazil. *Journal of Infections in Developing Countries* **4**: 533–45.

Fornadel, C. M., Norris, L. C. and Norris, D. E. (2010a) Centers for disease control light traps for monitoring *Anopheles arabiensis* human biting rates in an area with low vector density and high insecticide-treated bed net use. *American Journal of Tropical Medicine and Hygiene* **83**: 838–42.

Fornadel, C. M., Norris, L. C., Glass, G. E. and Norris, D. E. (2010b) Analysis of *Anopheles arabiensis* blood feeding behavior in southern Zambia during the two years after introduction of insecticide-treated bed nets. *American Journal of Tropical Medicine and Hygiene* **83**: 848–53.

Gimnig, J. E., Lindblade, K. A., Mount, D. L., et al. (2005) Laboratory wash resistance of long-lasting insecticidal nets. *Tropical Medicine and International Health* **10**: 1022–9.

Githinji, S., Herbst, S., Kistemann, T. and Noor, A. M. (2010) Mosquito nets in a rural area of Western Kenya: ownership, use and quality. *Malaria Journal* **9**: 250.

Gu, W. D. and Novak, R. J. (2009) Predicting the impact of insecticide-treated bed nets on malaria transmission: the devil is in the detail. *Mosquito Journal* **8**: 256.

Guyatt, H., Kinnear, J., Burini, M. and Snow, R. W. (2002) A comparative cost analysis of insecticide-treated nets and indoor residual spraying in highland Kenya. *Health Policy and Planning* **17**: 144–53.

Kayedi, M. H., Linesa, J. D. and Haghdoost, A. A. (2009) Evaluation of the wash resistance of three types of manufactured insecticidal nets in comparison to conventionally treated nets. *Acta Tropic* **111**: 192–6.

Kilian, A., Boulay, M., Koenker, H. and Lynch, M. (2010) How many mosquito nets are needed to achieve universal coverage? Recommendations for the quantification and allocation of long-lasting insecticidal nets for mass campaigns. *Malaria Journal* **9**: 330.

Kroeger, A., Ordonez-Gonzalez, J., Behrend, M. and Alvarez, G. (1999) Bed net impregnation for Chagas disease control: a new perspective. *Tropical Medicine and international Health* **4**: 194–8.

Kudom, A. A. and Mensah, B. A. (2010) The potential role of the educational system in addressing the effect of inadequate knowledge of mosquitoes on use of insecticide-treated nets in Ghana. *Malaria Journal* **9**: 256.

Lengeler, C. (2004) Insecticide-treated bed nets and curtains for preventing malaria (review). *The Cochrane Database of Systematic Reviews* 2. Art No. CD000363.pub2.

Lindblade, K. A., Dotson, E., Hawley, W. A., et al. (2005) Evaluation of long-lasting insecticidal nets after 2 years of household use. *Tropical Medicine and International Health* **10**: 1141–50.

Lindsay, S. W. and Gibson, M. E. (1988) Bed nets revisited – old idea, new angle. *Parasitology Today* **4**: 270-2.

Lines, J. D., Myamba, J. and Curtis, C. F. (1987) Experimental hut trials of permethrin-impregnated mosquito nets and eave curtains against malaria vectors in Tanzania. *Medical and Veterinary Entomology* **1**: 37-51.

Malima, R. C., Magesa, S. M., Tungu, P. K., et al. (2008) An experimental hut evaluation of Olyset® nets against anopheline mosquitoes after seven years use in Tanzanian villages. *Malaria Journal* **7**: 38.

Maxwell,C. A., Wakibara, J., Tho, S. and Curtis,C. F. (1998) Malaria-infective biting at different hours of the night. *Medical and Veterinary Entomology* **12**: 325-7.

Maxwell, C. A., Rwegoshora, R. T., Magesa, S. M. and Curtis, C. F. (2006) Comparison of coverage with insecticide-treated nets in a Tanzanian town and villages where nets and insecticide are either marketed or provided free of charge. *Malaria Journal* **5**: 44.

N'Guessan, R., Boko, P., Odjo, A., Chabi, J., Akogbeto, M. and Rowland, M. (2010a) Control of pyrethroid and DDT-resistant *Anopheles gambiae* by application of indoor residual spraying or mosquito nets treated with a long-lasting organophosphate insecticide, chlorpyrifos-methyl. *Malaria Journal* **9**: 44.

N'Guessan, R., Asidi, A., Boko, P., et al. (2010b) An experimental hut evaluation of PermaNet® 3.0, a deltamethrin piperonyl butoxide combination net, against pyrethroid-resistant *Anopheles gambiae* and *Culex quinquefasciatus* mosquitoes in southern Benin. *Transactions of the Royal Society of Tropical Medicine and Hygiene* **104**: 758–65.

Oxborough, R. M., Weira, V., Irish, S., et al. (2009) Is K-O Tab 1-2-3® long-lasting on non-polyester mosquito nets? *Acta Tropica* **112**: 49-53.

Pennetier, C., Chabi, J., Martin, T., et al. (2010) New protective battle-dress impregnated against mosquito vector bites. *Parasites and Vectors* **3**: 81.

Pettifor, A., Taylor, E., Nku, D., et al. (2009) Free distribution of insecticide treated bed nets to pregnant women in Kinshasa: an effective way to achieve 80% use by women and their newborns. *Tropical Medicine and International Health* **14**: 208.

Rossbach, B., Appel, K. E., Moss, K. G. and Letzel, S. (2009) Uptake of permethrin from impregnated clothing. *Toxicology Letters* **192**: 505.

Sharma, S. K., Upadhyay, A. K., Haque, M. A., et al. (2005) Village-scale evaluation of mosquito nets treated with tablet formulation of deltamethrin against malaria vectors. *Medical Veterinary Entomology* **19**: 286-92.

Skovmand, O. and Bosselmann, R. (2011) Strength of bed nets as function of denier, knitting pattern and polymer. *Malaria Journal* **10**: 87.

Smith, S. C., Joshi, U. B., Grabowsky, M., Selanikio, J., Nobiya, T. and Aapore, T. (2007) Evaluation of bed nets after 38 months of household use in Northwest Ghana. *American Journal of Tropical Medicine and Hygiene* **77 (Suppl 6)**: 243-8.

Soto, J., Medina, F., Dember, N. and Berman, J. (1995) Efficacy of permethrin-impregnated uniforms in the prevention of malaria and leishmaniasis in Colombian soldiers. *Clinical Infectious Diseases* **21**: 599-602.

Steketee, R. W. and Campbell, C. C. (2010) Impact of national malaria control scale-up programmes in Africa: magnitude and attribution of effects. *Malaria Journal* **9**: 299.

Tadei, W. P., Thatcher, B. D., Santos, J. M., Scarpassa, V. M., Rodrigues, I. B. and Rafael, M. S. (1998) Ecologic observations on anopheline vectors of malaria in the Brazilian Amazon. *American Journal of Tropical Medicine and Hygiene* **59**: 325-35.

Tungu, P., Magesa, S., Maxwell, C., et al. (2010) Evaluation of PermaNet 3.0, a deltamethrin-PBO combination net against *Anopheles gambiae* and pyrethroid resistant *Culex quinquefasciatus* mosquitoes: an experimental hut trial in Tanzania. *Malaria Journal* **9**: 21.

WHO (2002a) 'Impact of Olyset long lasting nets to control transmission of anthroponotic cutaneous leishmaniasis in Central Iran.' Final Technical Report. Project No: SGS04-76. WHO, Eastern Mediterranean Region.

WHO (2002b) 'Instructions for treatment and use of insecticide-treated mosquito nets. Use insecticide-treated mosquito nets to sleep in peace – and protect your health.' WHO/CDS/RBM/2002.41

WHO (2002c) 'Insecticide-treated mosquito net interventions. A manual for national control programme managers.' WHO/CDS/RBM/2002.45

WHO (2004) 'A generic risk assessment model for insecticide treatment of mosquito nets and their subsequent use.' WHO/CDS/WHOPES/GCDPP/2004.6; WHO/PCS/04.1

WHO (2005) 'Guidelines for laboratory and field testing of long-lasting insecticidal mosquito nets.' WHO/CDS/WHOPES/GCDPP/2005.11.

WHO (2007a) 'Long-lasting insecticidal nets for malaria prevention. Trial edition: a manual for malaria programme managers.' WHO, Geneva.

WHO (2007b) '*Insecticide-treated mosquito nets: a position statement.*' WHO, Geneva.

Yates, A., N'Guessan, R. N., Kaur, H., Akogbeto, M. and Rowland, M. (2005) Evaluation of KO Tab 1-2-3: a wash-resistant 'dip-it-yourself' insecticide formulation for long-lasting treatment of mosquito nets. *Malaria Journal* **4**: 52.

Zollner, G., Hoell, D., Hanafi, H. A., Richardson, J. H., Mukabana, R. and Coleman, R. E. (2007) Evaluation of novel long-lasting, insecticide-impregnated bed nets to control adult sand flies (Diptera: Phlebotominae) in human landing studies in Kenya and Egypt. *American Journal of Tropical Medicine and Hygiene* **77(5) Suppl. S**: 116–401.

Chapter 5
Larviciding

When Ross had shown that Anopheline mosquitoes were the vectors of malaria, he initiated considerable efforts to reduce the areas of water close to houses to minimise the breeding of larvae and he was later involved in campaigns in West Africa, the Suez Canal zone, Greece, Mauritius and Cyprus to increase drainage to reduce mosquito populations. Drainage schemes also played a key role in the first half of the twentieth century in eliminating malaria in many parts of Europe and the USA.

Apart from targeting Anopheline mosquitoes, drainage schemes have been very important in reducing populations of other mosquito vectors, notably *Aedes aegypti* to reduce transmission of dengue. Similarly, in the 1930s-1960s, massive ditching/impoundment programmes in coastal Florida reduced populations of salt marsh mosquitoes (*Aedes taeniorhynchus*), although not a disease vector. Where water cannot be drained, such as in irrigated crop areas, fish have been introduced, although this is not always feasible and in some cases their introduction may have an adverse effect on indigenous fish populations. Guppy (*Poecilia reticulata*) and Siamese fighting fish (*Betta splendens*) have been used in areas where vectors of dengue occur in Asia. Using fish that eat mosquito larvae and pupae has been promoted in some countries, by involving communities to assist their distribution and adoption.

Larviciding using insecticides is another method that can be used to control the immature stages of the mosquitoes, where water cannot be drained.

The spread of the exotic African malaria vector *Anopheles gambiae* in northeast Brazil in the 1930s led to a campaign of house spraying but the main effort in the eradication programme was larviciding using Paris green, a copper acetoarsenite, the success of which was due to a clearly defined and rigorously managed programme (Killeen et al., 2002; Soper and Wilson, 1943). However, in areas with intense transmission of malaria, it is usually impossible to treat every anopheline breeding site, so even if the population is

Integrated Vector Management: Controlling Vectors of Malaria and Other Insect Vector Borne Diseases, First Edition. Graham Matthews.

significantly reduced there may still be a high prevalence of malaria. Thus effective larval control is most feasible where breeding places are limited in number, easily recognisable and also easily accessible. Walker and Lynch (2007), reviewing larval control of *Anopheles* spp., suggested that larviciding is especially suitable in urban areas where larval habitats are limited as part of an integrated vector management programme, where other measures, such as indoor residual spraying (IRS) and insecticide treated nets (ITN) are also used, rather than as a stand-alone method. Larviciding is also more effective in areas where mosquito breeding is only during a short period and has been used in the USA and Europe to reduce populations of nuisance mosquitoes.

In addition to mosquitoes, water is the habitat of other vectors of disease, notably *Simulium* spp., the larvae of which need the oxygenated water in natural and man-made rapids along rivers. In a major international programme, *Simulium damnosum*, which transmits the microfiliariae of *Onchocerca volvulus* that causes river blindness, was controlled over an extensive area of West Africa from Senegal to Benin, by applying larvicides from aircraft (Lewis, 1974; Baker et al., 1985; Baldry et al., 1985). The Onchocerciasis Control Programme (OCP) initially applied the organophosphate temephos (Abate), which was very effective with minimal effects on non-target fauna. When resistance to temephos was detected, it was used in rotation with chlorphoxim, carbosulphan, *Bacillus thuringiensis israelensis* H14 (BTi), phoxim, pyraclofos, etofenprox and permethrin. In order of risks to non-target organisms, BTi caused the least damage, followed by temephos, chlorphoxim, pyraclofos, etofenprox, permethrin and carbosulfan, in increasing order of toxicity (Yaméogo et al., 1991, 1992, 2001, 2003; Léveque et al., 2003).

This programme extended over 20 years to interrupt the transmission of the disease, so it was no longer a public health problem and good arable land could be reclaimed (Samba, 1994). The main approach to controlling onchocerciasis is by taking the drug ivermectin annually for 15 years, the lifespan of the parasitic worm. This drug is donated free of charge by the Mectizan Donation Program and distributed under the African Programme for Onchocerciasis Control (APOC) and by various Non-Governmental Organisations (NGOs) such as Sight Savers, Lions International Sight First Programme, the Carter Foundation and the Helen Keller Foundation. However, there have been some indications of resistance to this drug (defined as individuals with microfilaria (mf) counts in skin of more than 10 mf/snip after nine or more rounds of ivermectin treatment) in Ghana (Dadzie et al., 2003; Awadzi et al., 2004) and a new drug, moxidectin, is being investigated for its potential to kill or sterilise the adult worms of *On. volvulus*. However, treatment of the disease does not remove the irritation of many bites by the flies, which has caused many people in rural areas to migrate to urban areas. Thus, control of the vector remains important to allow re-establishment of many farming areas and other developments.

This chapter deals with techniques used to apply larvicides in different situations.

Larvicide application

Any programme of larvicide application needs to be developed knowing the size, location and distribution of larval breeding sites. It is also important to consider the flight range of the vector species in order to determine the effectiveness or relevance of larviciding remote breeding sites, given the potentially huge economic cost of an intensive wide area larviciding programme.

Mosquito breeding site surveys are easily conducted in towns and other highly populated areas, whereas locating breeding sites in rural areas can be more difficult. In many situations there can be small transitory, but important sources of mosquitoes, which are not always easy to detect. Apart from swampy areas, irrigated fields, puddles, tyre tracks and streams (Figure 5.1), breeding sites include water containers (Figure 5.2) in areas without piped water supplies, vases of flowers (Figure 5.3), disused tyres (Figure 3.10), discarded bottles and other small receptacles of water, especially in relation to *Ae. aegypti* and *Aedes albopictus*, the principal vectors of dengue. Because of their association with urban areas and artificial containers, these two species are often referred to as 'domestic mosquitoes'. Urban areas often have campaigns to educate people in the importance of avoiding collection of water within their vicinity of their dwellings or to cover water containers with a lid. The use of community based schemes for dengue control has been discussed by Hahn et al. (2009); Toledo et al. (2007) and Vanlerberghe et al. (2009), and the cost effectiveness of control programmes involving the community by Baly et al. (2007).

Similarly for black fly control, the OCP carried out extensive mapping of all breeding sites in the dry and high water seasons, gaining access to some areas by helicopter.

The use of Global positioning and Geographical Information Systems (GPS/GIS) facilitates mapping the breeding sites of different vector species and retaining a record of where treatments are applied. It can also facilitate planning control operations. The rapid development of remote sensing techniques (aerial and satellite photographs, Doppler radar estimates of remote rainfall, multi-spectral imagery, etc.) has allowed the identification of potential breeding sites through habitat-type recognition and surface water identification on computer-based GIS programmes. This has significantly improved the ability to focus the labour-intensive mosquito breeding site surveys to those areas identified as most probable through the remote sensing techniques, allowing for more timely and targeted treatments.

Mosquito control

Although larval control has had considerable success, adoption of this technique to control *Anopheles* spp. in sub-Saharan Africa has not been actively promoted recently. Nevertheless, there is evidence from investigations at

(a)

(b)

Figure 5.1 (a) Small stream near village in Cameroon; (b) wet area in Zimbabwe (photo: Nigel Frazer-Evans).

(a)

(b)

Figure 5.2 (a) Water containers outside an African house; (b) traditional water pot outside a Malaysian house (photo: Chung gait Fee).

Mbita in western Kenya that microbial larvicides can substantially reduce human exposure to malaria and integrate effectively with the use of insecticide treated bed nets (Fillinger and Lindsay, 2006; Fillinger et al., 2009). These studies were preceded by weekly larval surveys over 20 months

Figure 5.3 Shrine alongside a Thai house with flowers in water, where mosquito larvae can live.

Figure 5.4 Field training in sampling for mosquito larvae (photo: Ulrike Fillinger).

to identify the availability of larval habitats, over 70% of which were man-made (Fillinger et al., 2004) (Figure 5.4) (Mukabana et al., 2006). Similar studies have been carried out in Benin (Becker, 2010) and The Gambia (Majambere et al., 2007).

Table 5.1 Recommendations for larviciding provided by WHO.

Insecticide	Formulation	Type	Dose g a.i./ha or as stated
Fuel oil	Solution		142–190 l/ha
	With spreading agent		19–47 l/ha
Bacillus thuringiensis	Water dispersible granules	Bio-insecticide	125–750 g of Product
Diflubenzuron	Granules or WP	IGR	25–100
Methoprene	EC	IGR	20–40
Novaluron	EC	IGR	10–100
Pyriproxyfen	Granules	IGR	5–10
Chlorpyrifos	EC	OP	11–25
Fenthion	EC; Granules	OP	22–112
Pirimiphos methyl	EC	OP	50–100
Spinosad	SC; Granules		20–50
Temephos	EC	OP	56–112

Oils

Larviciding with oils has in the past largely been through application of a petroleum oil, such as kerosene (paraffin) or diesel oil to the surface of mosquito breeding pools at the rate of approximately 150–200 l/ha, but reduced to 20–50 l/ha when a spreading agent is added. Control was improved by adding a surfactant (e.g. octoxinol) to increase the spread of the oil over the water surface. Problems occur when vegetation and other obstructions prevent adequate spread of the oil. Many have advocated using essential oils extracted from various plants as alternatives to petroleum products, but none have been recommended, as a number of other insecticide products have proved to be very effective larvicides. However, with resistance to insecticides used on bed nets, IRS and in agriculture, the importance of using oils has re-emerged. Djouaka et al. (2007) have reported bioassays examining use of kerosene against *An. gambiae* resistant to pyrethroids and determined that the lowest concentration that gave 100% mortality was $3,930 \times 10^{-3}$ µl/cm^2 and the LC$_{50}$ was $1,964 \times 10^{-3}$ µl/cm^2.

Some concerns have been expressed where an oil larvicide is applied aerially and spray drift occurs that may affect hatching of bird eggs (Hoffman et al., 2004).

Insecticides

A number of insecticides have been recommended for controlling mosquito larvae (Table 5.1). Temephos, a low toxicity organophosphate continues to be widely used, but in many situations, formulations of *B. thuringiensis* H-14 have been favoured. Federici (2010) has pointed out that development of recombinant DNA techniques has made it possible to develop new strains of B. thuringiensis, which are 10-fold more toxic than wild types currently used in commercial formulations. Bacillus sphaericus has also been applied as a larvicide, although its efficacy does vary between target species. Similarly, insect growth regulators have been preferred, except in areas with crustaceans (i.e. crab, shrimp, crawfish or lobsters), as they have the potential to cause adverse effects on these non-targets. This is particularly true in the case of the chitin-synthesis inhibitors, diflubenzuron and novaluron. A relatively new bacterial-toxin-based insecticide, spinosad, is now recommended as a larvicide (Hertlein et al., 2010).

WHO has published guidelines on risk assessment of larvicides (WHO, 2010a) and for evaluating and field testing of insecticides as larvicides (WHO, 2005). In contrast to the situation with insecticides to control adult vectors, there is a greater range of larvicides to allow a rotation of active ingredients to minimise selection of resistant populations.

Application of mosquito larvicides

Knapsack spraying

The simplest equipment for larvicide application is usually a small lever-operated knapsack sprayer (Figure 5.5a), often fitted with a cone nozzle. Ideally the spray is applied at low pressure with large droplets allowing the oil or insecticide formulation to spread across the water surface. Use of large droplets avoids spray drifting downwind, so it can be directed to quite small pools. If a compression sprayer is used, it should be fitted with a constant flow valve at a low pressure (1 or 1.5 bar) to ensure that large droplets are applied.

This type of portable equipment is suitable for applying water dispersible granules to extensive areas of stagnant water with little emergent or floating vegetation, which might prevent the spray from reaching the water surface. In Benin, after precise mapping of each significant breeding site in one district of Cotonou, a team of two people equipped with a knapsack sprayer treated the areas with B. thuringiensis (Vectobac) or B. sphaericus (Vectolex) water dispersible granules mixed in 5 litres of water and applied at 0.25–0.5 kg/ha (Becker, 2010).

Motorised equipment

Truck mounted or portable motorised hydraulic sprayers have been used, often with off-set nozzles to apply a larvicide alongside roads where water is not adequately drained. Off-set or boomless nozzles can be used to project the spray over a wide swath, depending on the pressure and output of the selected nozzle.

Recent trials with air-assisted ultra-low volume (ULV) portable and truck mounted sprayers using rotary atomisers has proved successful with application

(a)

(b)

Figure 5.5 (a) Applying larvicide with a knapsack sprayer in the Gambia (from Majambere et al., 2007); (b) application of dry corn granules to vegetated wet area in the Gambia (from Majambere et al., 2007).

of Bti (*B. thuringiensis israeliensis*) as small 'driftable' droplets (30-80 μm diameter) for 'broadcast' larvicide application in areas with multiple isolated pools of water where mosquitoes can breed, such as residential homes, wetlands and other areas where it is often difficult to apply larvicides directly to water.

Although thermal fogging is used primarily as a space treatment, fogging *B. thuringiensis* (Bti) formulations diluted in water has also been used to control larvae (Seleena et al., 2001). By selecting a higher flow rate to avoid the risk of clogging the nozzle, larger droplets are produced that sediment on water surfaces. This technique has not been generally recommended as some of the fog will disperse downwind, but the larger droplets will deposit over a relatively short distance (~3-4 m) from the operator. Yap et al. (2002), using a water dispersible granule formulation, showed good larvicidal and residual activity in small containers of water. Chung et al. (2001) used a mixture of insecticides (pirimiphos methyl and Bti) to control adults and larvae simultaneously.

Aerial application for mosquito control

Aircraft have been used to apply both sprays and granules for treating mosquito breeding sites.

Application of aerial sprays

Agricultural areas, such as irrigated rice fields and natural wetlands close to villages and towns are often significant mosquito production sites (Figure 5.6). Aerial application of a larvicide is often the most appropriate means of controlling these populations (Figure 5.7a). When the aircraft is not fitted with a GPS, it is still possible to use a hand carried GPS unit with a laptop and Google map to record the flight path during a spraying operation (Figure 5.7b).

In field trials comparing two droplet size distributions (Volume Median Diameter (VMD) 67 μm vs. 225 μm) achieved with AU4000 rotary atomiser and CP nozzle applying four different application rates of *B. thuringiensis* (Vectobac 12AS), the smaller droplets gave higher mortality at the lower deposit rates but there were more droplets per unit area (Mickle, 2004) (Figure 5.8) where there was no foliage covering the ponds. However, with downwind movement of smaller droplets, flight paths need to be offset upwind.

Later trials, using AU5000 atomisers set to produce a VMD of 135 μm over water sheltered by thatch, grass and reeds, showed that little spray penetrated thick thatch, while grass captured 25% of the larvicide (Mickle, 2001). Penetration was better through reeds by droplets smaller than 150 μm in contrast to grass, where penetration was better with the larger droplets, presumably due to differences in the air flow within the vegetation. Subsequent studies comparing sprays with granule formulations showed that penetration of different canopy types was better with granule application, averaging 70% compared with 45% with sprays (Mickle, 2004). As noted with reeds, better penetration of larvicide sprays was through tall canopies and low ground cover with smaller droplets. Generally aerial larviciding uses a directed spray of large droplets with a negligible volume in fine droplets. Total application

(a)

(b)

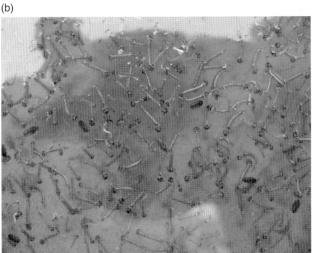

Figure 5.6 (a) Pool with mosquito larvae, Greece; (b) close-up showing high density of larvae (photos: Mark Latham).

volumes (active product formulation plus water as diluent) vary widely between programmes, the determining factors being target habitat, aircraft type and nozzle-type/droplet size. Spray volumes as low as 5 litres per hectare (Figure 5.7a) have been used, usually over rice fields, but rates as high as 50 litres per hectare and as low 1 litre per hectare have been utilised.

Application of granules

Granule formulations have been preferred for many larvicide control operations, as the solid particles fall through vegetation and thus penetrate down to the water. Granules designed to float can achieve greater contact

(a)

(b)

Figure 5.7   (a) Helicopter applying Bti larvicide at 5l/ha in Greece; (b) GPS track overlaid on a Google map (photos: Mark Latham).

with surface-feeding Anopheline mosquitoes. Granule application is suitable for most habitats under different environmental conditions and can be readily applied by communities with minimum training. Typical application rates for formulated granules are between 0.25 and 1g/m^2.

Ground application

A number of small applicators, such as a large type of 'pepper pot' have been used to apply granules to water surfaces. Small areas may be treated using a gloved hand (Figure 5.5b) In Asia, most houses without piped water have several types of water-storage containers, that vary from plastic pails, large glazed clay jars, metal drums to large tanks ranging from 20–2,000-litre capacity, all of

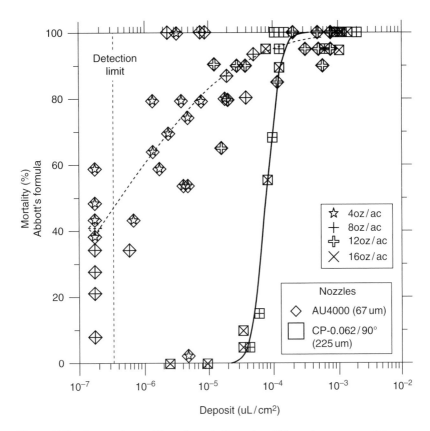

Figure 5.8 Comparison of larval mortality using different spray nozzles providing large and small droplets (data courtesy of Bob Mickle).

which can support populations of *Ae. aegypti*. To treat potable water requires a very low toxicity insecticide, which does not leave an unacceptable taste in the water. Trials by Bang and Tonn (1969a), which showed that 1ppm temephos applied as 1% sand granules provided complete control and protection against *Ae. aegypti* re-infestation for 20 weeks in water-storage jars. This led to the use of granules as part of a dengue control programme (Gratz, 1993).

More recently, Thavara et al. (2004) reported on use of 1% temephos on zoelite granules and showed that larvae were controlled for at least 2 months, but numbers remained low for up to 5 months. The problem is to keep such containers covered to prevent further oviposition. Some larvicides have been marketed as briquette/tablet formulations, particularly for container treatments, water collection barrels, cisterns and storm drain/catch-basins, for which specific slow release tablet or briquette formulations are designed to treat up to 10 m².

Larger areas of shallow water may be treated using a 'horn seeder', which consists of a tapered discharge tube with a variable opening that is fixed at the lowest point of a shoulder-slung bag, fitted with a zipped opening for filling. A swath of up to 7 m can be achieved by swinging the discharge tube in a

Figure 5.9 Horn seeder for applying granules to small areas (courtesy of Bob Mickle).

figure-of-eight pattern (Figure 5.9). Sutherland and White (1978) reported on trials which demonstrated that many factors influenced granular application using a Horn-seeder®. These included the setting of the seeder, weight and particle size of the granules, the height and walking speed of the applicator, the angle of the horn, the arc, pattern and force of swing, the number of swings per unit distance, and the terrain. Although it is possible to calibrate the seeder to apply a measured dose, the situation in the breeding site with vegetation and other obstructions will prevent a uniform walking speed and swinging action.

Alternatively, it is possible to project granules over water using a motorised mist blower adapted for applying dry formulations. The ideal equipment has a hopper with a sloping floor that enables the dry granules to fall towards the outlet connected to a delivery tube through which a high velocity air stream is projected from a centrifugal fan. A metering system controls the flow of granules into the air stream. Equipment with smaller manually operated fans can also be used, but the distance over which the granules are projected is limited.

Aerial application

Aerial granule applicators consist of a hopper (or two hoppers arranged on each side of a helicopter) from which the granules are fed through a metering system, which can be controlled electronically by the pilot (Figures 5.10 a, b, c, d, e and f).

In assessing the impact of *B. thuringiensis* granules on anopheline populations in rice fields, Lacey and Inman (1985) used a Grumman G-164A Ag

Cat airplane fitted with a 'Transland Deep Throat' high volume granule spreader to treat 15.2-m swaths from a height of 4–5 m and flying at an approximate air speed of 160 km/hr. This technique delivered 11.2 kg/ha with a single pass along each swath, but a lower rate (5.6 kg/ha), achieved by diluting the granules with an inert granule, was as effective as the higher rate tested.

Mickle (2004) reported trials using a Robinson Model 480 spreader system to determine the optimal swath width for the Hughes 500 helicopter flying at

(a)

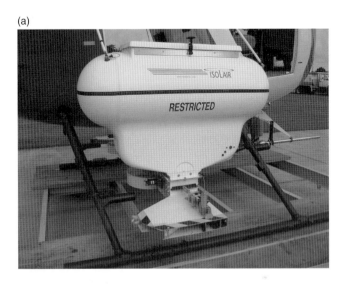

(b)

Figure 5.10 (a) Granule equipment on a helicopter; (b) close-up of dispenser; (c) view of granule hopper. A hopper is fitted on both sides of the helicopter; (d) metering system at bottom of the hopper; (e) different collecting trays used in an experiment to assess the swath width from a granule applicator; (f) release of granules over a mosquito breeding site, Florida (photos: Mark Latham).

(c)

(d)

Figure 5.10 *(Continued)*

(e)

(f)

Figure 5.10   *(Continued)*

23 m height and to characterise in-swath variability of both deposit (kg/ha) and distribution of granules (number of granules/m²) of two types of granule, one being based on a corn cob granule. The swath was assessed by collecting granules in 5-US-gallon buckets (opening area 0.064 m²) placed at 0.6 m intervals across the flight path (Mickle, 2002) (Figure 5.10e).

In the field, samples were transferred to holding tubes and subsequently weighed to 0.001 g accuracy. The distribution of granules across a swath is similar to sprays with lighter or smaller granules being carried further from the flight path, thus peak deposit is affected by wind speed. Thus, with low wind speeds, peak deposit was generally less than 6 m downwind of the flight line. As winds approached 12 km/h, the flight path needed to be off-set by one operational swath width (20 m) to deposit granules on a small area, otherwise control would be poor.

The cost of aerial application of larvicides is significantly higher, especially granule applications (up to $80/ha) than for apply adulticide sprays (<$5/ha), due to the higher volumes applied and narrower track spacings needed.

Monitoring

Before treating a body of water with a larvicide, it is important to take samples to assess the population and its distribution within an area. A long handled white dipper that can collect at least 100 ml of water is often used to collect mosquito larvae from surface water (Figure 5.4). Differences in the behaviour of larvae, time of day and weather conditions can affect sampling and it is difficult to relate numbers collected to absolute larval populations (Service, 1993; Skovmand et al., 2000). Similar sampling, after the application of larvicide, is important to assess the impact of the control measures. Mulla et al. (1971) developed a simple formula that took into account the population of the target species in sprayed and unsprayed areas, pre- and post-treatment.

The percent control or reduction (R) is $100 - [(C_1/T_1) \times (T_2/C_2) \times 100]$. 1 and 2 refer to pre- and post-treatment counts in C untreated and T treated areas. The formula assumes that counts in individual traps are independent measures of mosquito abundance and changes in the ratio between control and treated areas were due to the treatment.

Black flies

Black fly larvae live in highly oxygenated water, so populations are largely restricted to rivers with naturally occurring rapids, but the development of hydro-electric dams with a spillway to enable excess water to by-pass the dam has created other breeding sites (Figure 5.11). Other man-made structures such as bridges can also affect the water flow. The nuisance problem caused by the biting of black flies, in addition to causing 'river blindness', has increased. This has occurred particularly in Cameroon in areas along the River Sanaga and its main tributary, the Mbam, where the presence of hydro-electric dams at Edea and Songloulou have added to the naturally occurring rapids, such as the Kikot Falls (Boussinesq et al., 1992; Macé et al., 1997). Medical surveys using skin snips have shown that at least 50% of the population can be infected with *On. volvulus*. According to Remme et al. (1986), some people may have 100 micro-filariae per skin snip and blindness rates may exceed 5%

(a)

(b)

Figure 5.11 (a) Hydro-electric dam; (b) dry spillway, which creates white water when dam overflows.

Table 5.2 Insecticides applied as larvicides against *Simulium* spp.

Insecticide	Formulation and concentration g/l	Type	Dose l/m³/s
Bacillus thuringiensis	*WG*	*Bio insecticide*	*0.54-0.72*
Carbosulfan	EC 250	C	0.12
Phoxim	EC 500	OP	0.16
Pyraclofos	EC 500	OP	0.12
Temephos	EC 200	OP	0.15-0.3
Permethrin	EC 200	PY	0.045
Etofenprox	EC 200	PY	0.06

(Pion et al., 2002). People are so bitten by the flies, especially near the river, that they are unable to work, even when their body is well covered to reduce flies landing on exposed skin. Ideally there should be a Sanaga River Authority to integrate the control of black flies and mosquitoes throughout the river area, so that both industrial and agricultural development can take place.

Insecticides

The choice of insecticide needs to be decided after small-scale tests, as resistance has been detected after prolonged use of one insecticide. Jamnback and Frempong-Boadu (1966) documented early tests with DDT and temephos. Kurtak et al. (1987) have also reported on screening for insecticides to control *Simulium*. Apart from a low mammalian toxicity, the key factor as the insecticide is added to water is the impact on fish and other non-target organisms in the river. Currently the insecticides recommended for use as larvicides, are shown in Table 5.2. These would be expected to also have an impact on mosquito larvae downstream of any area of rapids treated for black fly control.

Aerial application of larvicides for black fly control

Although some treatments were made using boats, the OCP programme relied on aerial application of larvicides (Davies et al., 1978). The area covered was 765,000 km² by 1979, with 18,000 km of river being treated to protect a population of 16.5 million. Initially weekly treatments were made at known breeding sites, but as there was more knowledge of the *Simulium* population, treatments were more selective. A mixed fleet of aircraft (fixed wing and helicopter) were required due the differences in the sites and distances (Parker, 1975).

The dose of insecticide has to be designed to have an effect as the water passes through the rapids. Thus the amount of insecticide will be affected by the volume of water being discharged through the larval habitat. The discharge rate is measured in cubic metres per second (m³/s) (Hill, 1959).

Flowmeters, such as 'Flo-Mate Portable flowmeter (Marsh-McBirney Inc., USA.) can be used to measure the flow of water, but if a suitable meter is not available, the flow can be calculated from a measurement of the cross-section of the river and assessing the velocity mid-stream. On small streams this can be done by marking out a 10 m section of the stream without excessive turbulence, and then measuring the width and depth of the stream at several locations along this section. Having determined the cross-sectional area, a float is placed in the stream a few metres before the upstream marker and by using a stopwatch, the time it takes to reach the downstream marker is measured. The velocity should be measured several times and an average value used in the calculation of stream discharge–velocity (m/s) × average depth (m) × average width (m) = m^3/s (Skovmand et al., 2000). As an example if the average velocity is 0.65 m/s, average depth is 0.45 m and average width is 5 m, then the discharge rate is 1.46 m^3/s. Stream width has also provided a means of assessing the dosage to be applied (Undeen and Malloy, 1996).

The amount of insecticide used by the OCP project was calculated to be between 1 part insecticide to 20 million parts of water and 1 part to 1 million parts of water over a period of 10–30 minutes at a fixed site (Walsh, 1983). Thus, when applying temephos, at rates of 0.05 ppm to 0.1 ppm for 10 minutes, the insecticide is taken down river for about 45 km. The distance the larvicide has an effect depends on the flow of water and characteristics of the river. Slow moving water or the presence of pools reduces the distance downstream that an application is effective. Particulate formulations are only effective over a relatively short distance. Formulations of Bti are normally only applied when water levels are low, to minimise the dose required.

Trials on small rivers have indicated that it is better to apply the insecticide over a short period of time so that the larvae are exposed to a higher concentration than if the dosage was applied to the water over a longer period. Thus 10 mg/l of a Bti formulation per 1 minute was significantly more effective than 0.5 mg/l discharged per minute over 20 minutes (Lacey and Undeen, 1984).

A novel technique of application was developed – the rapid release ('Vide Vite') system, in which a pre-determined dose of the larvicide (Abate EC) was dropped upstream of a rapid (Figure 5.12a), so that the temephos formulation mixed with the water in the river, spreading both laterally across the river and downwards within the flow of water, so that on reaching the rapids, all the water contained sufficient temephos (Lee et al., 1975). The equipment (Figure 5.12b) was mounted below the aircraft with its long axis parallel with the line of flight. In the initial trials, with the equipment fitted to a Pilatus Porter aircraft, the pilot was able to place the insecticide accurately, some-times within a metre, at more than 50 l/s (Parker, 1975). This technique did depend considerably on the quality of the formulation, especially its specific gravity, so that it mixed well with the water before reaching the rapids.

However, when resistance to temephos was detected, they applied *B. thuringiensis israelensis* Berliner serotype H-14 and as this was a particulate formulation and did not spread well in the water, it had to be applied more conventionally by spraying across the width of the river. This was achieved

(a)

(b)　　　　Schematic layout of rapid release system

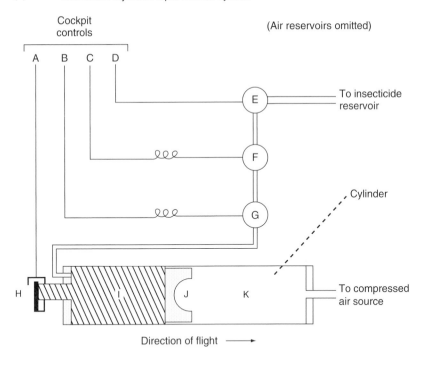

KEY　A. Gate opening/closing control　　　G. Solenoid valve
　　　B. ON/OFF solenoid value control　　H. Outlet gate
　　　C. Flowmeter totalizer　　　　　　　　I. Insecticide to be emitted
　　　D. Insecticide liquid pressure gauge　J. Piston
　　　E. Insecticide pump　　　　　　　　　K. Compressed air space
　　　F. Flow meter

Figure 5.12 (a) Helicopter applying larvicide for Simulium larval control during the WHO Onchocerciasis Control Programme (OCP); (b) OCP Vide Vite spray apparatus (from Parker, 1975).

(a)

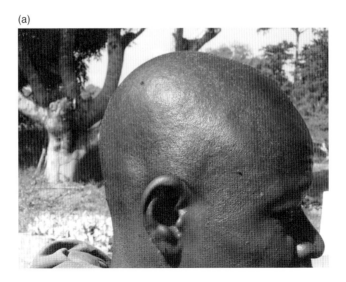

(b)

Figure 5.13   (a) Black fly on fisherman's head, Sanaga River, Cameroon;
(b) examining villagers and taking skin snips to detect *Onchocerca*; (c) villager
with river blindness; (d) wife of villager guiding her blind husband; (e) skin
discoloration due to scratching legs following black fly bites; (f) collecting
black fly adults.

(c)

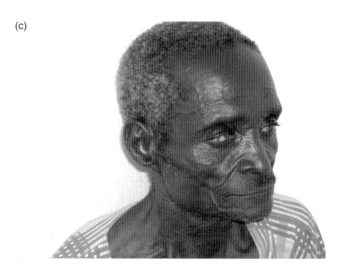

(d)

(e)

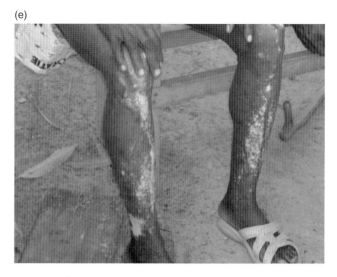

Figure 5.13 (*Continued*)

(f)

Figure 5.13 *(Continued)*

by fitting a control system to a Simplex tank so that the required dose could be delivered to a boom fitted with a series of nozzles, each controlled by a solenoid valve. Rather than the larvicide being sprayed, it was dribbled out with the number of nozzles and their size determined in relation to the width and flow of the river. During the application, the pilot would fly across the river if possible, but the actual flight path depended on the width of the river and the vegetation/trees along the banks. The Simplex system was applicable to helicopters, whereas on fixed-wing aircraft the standard aircraft spray system was adapted with a different pump, nozzles and control system.

Boat application

The cost of maintaining an aerial application is very high, so when it was decided to treat the population with ivermectin, efforts on controlling the vector declined, but regardless of being the disease vector, the biting by large numbers of black flies is intolerable in some riverine areas (Figure 5.13), so localised treatment of areas severely affected by the flies has been done using a boat (pirogue) – the Simulium Pirogue Application Technique (SPATE) (Figure 5.14c) (Matthews et al., 2009). By fitting an outboard motor to the boat, it was possible to cross a wide river during a period of application, so that the whole width of the river was treated. In one example, temephos was applied at a dosage of 0.06–0.12 l of Abate 500 EC per m^3 per second flow of the river through four large cone nozzles without a swirl plate at 2.5 l/min/nozzle to deliver 50 litres of insecticide in 5 minutes. It is usual to apply 3 applications at 7–10 day intervals to reduce the black fly population, but the situation has to be monitored and further treatments made if the number of flies exceeds a tolerable number. The technique is thus easily carried out by a small team of trained people living near the rapids.

Applications in small streams

In treating smaller streams too shallow for a boat, if access to the bank is possible upstream of a black fly breeding site, it should be possible to treat the width of the river by using a motorised hydraulic knapsack sprayer fitted with a boomless nozzle. The aim of this type of nozzle is to create a wide swath across the stream. On very narrow streams it has been possible to dispense the larvicide directly into the river from a single point using a watering can (Lacey and Undeen, 1984), although care is needed to ensure that it does get mixed across the full width of a stream.

Monitoring

Various techniques have been used to assess the impact of larvicide application. One technique is to place artificial targets with a defined surface area, such as plastic ribbons, into the fast flowing water before and after treatment. Artificial targets are normally left for a defined period, for example, 24 hours to allow larvae to attach to them. Counts are made of the larvae attached to the ribbons placed before and after treatment to show a reduction in larvae.

Alternatively, larvae attached to targets in untreated and treated water are transferred to a bowls of water, which are constantly aerated over 24 hours or longer to assess mortality due to the treatment.

An important assessment is to sample the adult black fly population. This has been done with Perspex sheets covered in glue (Figure 5.14d and Figure 5.15) (Bellec and Hebrard, 1976).

(a)

Figure 5.14 (a) White water providing breeding site for black fly larvae, Sanaga River, Cameroon; (b) larvae on vegetation taken from rapids (photo: Hans Dobson); (c) treating a river with larvicide from a boat (photo: Didier Baleguel); (d) treating Perspex sheet with glue to sample adult black flies following treatment of the river.

(b)

(c)

(d)

Figure 5.14 *(Continued)*

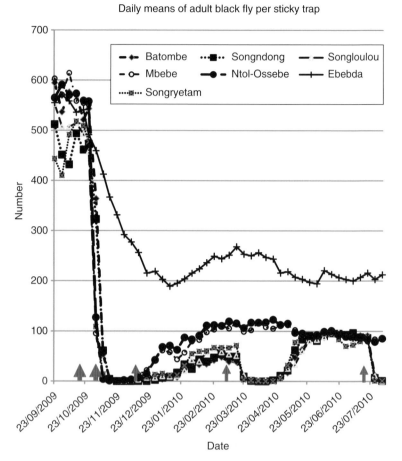

Figure 5.15 Average counts of black fly per day on sticky Perspex traps at different sites along the Sanaga River, Cameroon, during 2009/10, following larvicide treatments upstream of the hydro-electric dams and also the Kikot rapids. Three applications at 7 day interval at each treatment period decided following data from stick traps. Ebebda was upstream of all treated areas.

Conclusion

Larviciding is a crucial part of integrated vector management and from the pesticide aspect provides different modes of action that may be invaluable in maintaining vector populations at low levels. It is also considered important, as it can reduce vector populations before the adult vectors can bite or transmit disease. Construction of dams should also take into account the design of spillways to minimise the creation of suitable habitats for insect larvae.

References

Awadzi, K., Attah, S. K., Addy, E. T., et al. (2004) Thirty-month follow=up of sub-optimal response to multiple treatments with ivermectin in two onchocerciasis-endemic foci in Ghana. *Transactions of Tropical Medicine and Parasitology* **98**: 359-70.

Baker, R. H. A., Baldry, D. A. T., Zerbo, D. G., Pleszak, F. C., Boakye, D. and Wilson, M. (1985) Measures aimed at controlling the invasion of *Simulium-damnosum theobald* s.l. (Diptera, Simuliidae) into the Onchocerciasis Control Program area. 2. Experimental aerial larviciding in the Sankarani and Fie basins of Eastern Guinea in 1984 and 1985. *Tropical Pest Management* **32**: 148-61.

Baldry, D. A. T., Zerbo, D. G., Baker, R. H. A., Walsh, J. F. and Pleszak, F. C. (1985) Measures aimed at controlling the invasion of *Simulium-damnosum theobald* s.l. (Diptera, simuliidae) into the Onchocerciasis Control Program area. 1. Experimental aerial larviciding in the upper Sassandra basin of Southeastern Guinea in 1985. *Tropical Pest Management* **31**: 255-63.

Baly, A., Toledo, M. E., Boelaert, M., et al. (2007) Cost effectiveness of *Aedes aegypti* control programmes: participatory versus vertical. *Transactions of the Royal Society of Tropical Medicine and Hygiene* **101**: 578-86.

Bang, Y. H. and Tonn, R. J. (1969a) Evaluation of 1% Abate sand granules for the control of *Aedes aegypti* larvae in potable water. Unpublished document WHO/VBC/69.121.

Becker, N. (2010) The Rhine larviciding program and its application to vector control. In Atkinson, P. W. (ed.), *Vector Biology, Ecology and Control*. Springer, Dordrecht.

Bellec., C. and Hebrard, G. (1976) Captures d'adultes de Simuliidae, en particulier de *Similium damnosum* Theobald, 1903, a l'aide de pieges d'inception: les pieges-vitres. Paris: Office de la Recherche Scientifique et Technique Outre-Mer, OCCGE/ORSTOM; No. 17/Oncho/Rap/76, mimeographed document.

Boussinesq, M., Demanga-Ngangue, R. P., Lele, D., Cot, S. and Chippaux, J. P. (1992) Etude clinique et parasitologique de l'onchocercose dans huit villages de la vallée du Mbam (Province du Centre, Cameroun). *Bulletin de Liaison et de Documentation de l'O.C.E.A.C.* **100**: 26-31.

Chung, Y. K., Lam-Phua, S. G., Chua, Y. T. and Yatiman, R. (2001) Evaluation of biological and chemical insecticide mixture against *Aedes aegypti* larvae and adults by thermal fogging in Singapore. *Medical and Veterinary Entomology* **15**: 321-7.

Dadzie, Y., Neira, M. and Hopkins, D. (2003) Final report of the Conference on the eradicability of Onchocerciasis. *Filaria Journal* **2(2)**: + 141 pp.

Davies, J. B., Le Berre, R., Walsh, J. F. and Cliff, B. (1978) Onchocerciasis and *Simulium* control in the Volta River Basin. *Mosquito News* **38**: 466-72.

Djouaka, R. F., Bakare, A. A., Bankole, H. S., et al. (2007) Quantification of the efficiency of treatment of *Anopheles gambiae* breeding sites with petroleum products by local communities in areas of insecticide resistance in the Republic of Benin. *Malaria Journal* **6**: 56.

Federici, B. A. (2010) Recombinant bacterial larvicides for control of important mosquito vectors of malaria. In Atkinson, P. W. (ed.), *Vector Biology, Ecology and Control*. Springer, Dordrecht.

Fillinger, U. and Lindsay, S. W. (2006) Suppression of exposure to malaria vectors by an order of magnitude using microbial larvicides in rural Kenya. *Tropical Medicine and International Health* **11**: 1629-42.

Fillinger, U., Sonye, G., Killeen, G. F., Knols, B. G. J. and Becker N. (2004) The practical importance of permanent and semi-permanent habitats for controlling aquatic stages of *Anopheles gambiae sensu lato* mosquitoes: operational observations from a rural town in western Kenya. *Tropical Medicine and International Health* **9**: 1274-89.

Fillinger, U., Ndenga, B., Githeko, A. and Lindsay, S. W. (2009) Integrated malaria vector control with microbial larvicides and insecticide-treated nets in western Kenya: a controlled trial. *Bulletin of the World Health Organisation* **87**: 655-65.

Gratz, N. G. (1993) Lessons of *Aedes aegypti* control in Thailand. *Medical and Veterinary Entomology* **7**: 1-10.

Hanh, T. T., Hill, P. S., Kay, B. H. and Quy, T. M (2009) Development of a framework for evaluating the sustainability of community-based dengue control projects. *Journal of the American Mosquito Control Association* **80**: 312-18.

Hertlein, M. B., Mavrotas, C., Jousseaume, C., et al. (2010) A review of spinosad as a natural product for larval mosquito control. *Journal of the American Mosquito Control Association* **26**: 67-87.

Hill, G. (1959) Methods of water-flow measurement applicable to *Simulium* control. *Bulletin of the World Health Organisation* **21**: 201-5.

Hoffman, D. J., Albers, P. H., Melancon, M. J. and Miles, A. K. (2004) Effects of the mosquito larvicide GB-1111 on bird eggs. *Environmental Pollution* **127**: 353-8.

Jamnback, H. and Frempong-Boadu, J. (1966) Testing black fly larvicides in the laboratory and streams. *Bulletin of the World Health Organisation* **34**: 405-21.

Killeen, G. F., Fillinger, U., Kiche, I., Gouagna, L. C. and Knols, B. G. J. (2002) Eradication of *Anopheles gambiae* from Brazil: lessons for malaria control in Africa? *The Lancet Infectious Diseases* **2**: 618-27.

Kurtak, D., Jamnback, H., Meyer, R., Ocran, M. and Renaud, P. (1987) Evaluation of larvicides for the control of *Simulium damnosum* s.l. (Diptera: Simuliidae) in West Africa. *Journal of the American Mosquito Control Association* **3**: 201-10.

Lacey, L. A. and Inman, A. (1985) Efficacy of granular formulations of *Bacillus thuringiensis* (H-14) for the control of *Anopheles* larvae in rice fields. *Journal of the American Mosquito Control Association* **1**: 38-42.

Lacey, L. A. and Undeen, A. H. (1984) Effect of formulation, concentration and application time on the efficacy of *Bacillus thuringiensis* (H-14) against black fly (Diptera: Simuliidae) larvae under natural conditions. *Journal of Economic Entomology* **77**: 412-18.

Lee, C. W., Parker, J. D., Philippon, B. and Baldry, D. A. T. (1975) Prototype rapid release system for the aerial application of larvicide to control *Simulium damnosum* Theo. *PANS* **21**: 92-102.

Léveque, C., Hougard, J. M., Resh, V., Statzner, B. and Yameogo, L. (2003) Freshwater ecology and biodiversity in the tropics: what did we learn from 30 years of onchocerciasis control and the associated biomonitoring of West African rivers? *Hydrobiologia* **500**: 23-49.

Lewis, D. J. (1974) 'A review of aerial control of black flies (Simuliidae) with reference to tropical Africa.' WHO/VBC/74.471.

Macé, J. M., Boussinesq, M., Ngoumou, P., Enyegue Oye, J., Koéranga, A. and Godin, C. (1997) Country-wide rapid epidemiological mapping of onchocerciasis (REMO) in Cameroon. *Annals of Tropical Medicine and Parasitology* **91**: 379-91.

Majambere, S., Lindsay, S. W., Green, C., Kandeh, B. and Fillinger, U. (2007) Microbial larvicides for malaria control in The Gambia. *Malaria Journal* **6**: 76.

Matthews, G. A., Dobson, H. M., Nkot, P. B., Wiles, T. L. and Birchmore, M. (2009) Preliminary examination of integrated vector management in a tropical rain forest area of Cameroon. *Transactions of the Royal Society of Tropical Medicine and Hygiene* **103**: 1098-104.

Mickle, R. E. (2001) 'Characterization and Penetration Studies for Vectobac 12AS.' REMSpC Report 2001-15.

Mickle, R. E. (2002) Lines of flight – swath characterization and block variability. *Wing Beats* **13(4)**: 22-5.

Mickle, R. E. (2004) 'Prerequisites and equipment availability for an efficient adulticide treatment program for mosquito control.' REMSpC Report 2004-03.

Mukabana, W. R., Kannady, K., Kiam, G. M., et al. (2006) Ecologists can enable communities to implement malaria vector control in Africa. *Malaria Journal* **5**: 9.

Mulla, M. S., Norland, R. L., Fanara, D. M., Darwezeh, H. A. and Mc Kean, D. W. (1971) Control of chironomid midges in recreational lakes. *Journal of Economic Entomology* **64**: 300-7.

Parker, J. D. (1975) The use of aircraft in the WHO Onchocerciasis Programme. *Proceedings of the Fifth International Agricultural Aviation Congress*. 127-36.

Pion, S. D. S., Kamgno, J., Demanga-Ngangue and Boussinesq, M. (2002) Excess mortality associated with blindness in the onchocerciasis focus of the Mbam Valley, Cameroon. *Annals of Tropical Medicine and Parasitology* **96**: 181-9.

Remme, J., Ba, O., Dadzie, K. Y. and Karam, M. (1986) A force-of-infection model for onchocerciasis and its applications in the epidemiological evaluation of the Onchocerciasis Control Programme in the Volta River basin area. *Bulletin of the World Health Organization* **45**: 517-20.

Samba, E. M. (1994) *The Onchocerciasis Control Programme in West Africa*. WHO, Geneva.

Seleena, P., Lee, H. L. and Chiang, Y. F. (2001) Thermal application of *Bacillus thuringiensis* serovar *israelensis* for dengue control. *Journal of Vector Ecology* **26**: 110-13.

Service, M. W. (1993) *Mosquito Ecology – Field Sampling Methods*. Elsevier Applied Science. London.

Skovmand, O., Kerwin, J. and Lacey, L. A. (2000) Microbial control of mosquitoes and black flies. In: Lacey, L. A. and Kaya, H. K. (eds), *Field Manual of Techniques in Invertebrate Pathology*. Kluwer, Dordrecht.

Soper, F. L. and Wilson, D. B. (1943) *Anopheles gambiae in Brazil 1930 to 1940*. Rockefeller Foundation, New York.

Sutherland, D. J. and White, D. J. (1978) Distribution pattern of temephos granules applied by the Horn seeder. *Mosquito News* **38**: 127-31.

Thavara, U., Tawatsin, A., Kong-Ngamsuk, W. and Mulla, M. S. (2004) Efficacy and longevity of a new formulation of temephos larvicide tested in village-scale trials against larval *Aedes aegypti* in water storage containers. *Journal of the American Mosquito Control Association* **20**: 176-82.

Toledo, M. E., Vanlerberghe, V., Balya, A., et al. (2007) Towards active community participation in dengue vector control: results from action research in Santiago de Cuba, Cuba. *Transactions of the Royal Society of Tropical Medicine and Hygiene* **101**: 56-63.

Undeen A. H. and Malloy, D. P. (1996) Using stream width for determining the dosage rates of *Bacillus thuringiensis* Var *Israelensis* for larval black fly (Diptera: Simuliide) control. *Journal of the American Mosquito Control Association* **12**: 312-15.

Vanlerberghe, V., Toledo, M. E., Rodriguez, M., et al. (2009) Community involvement in dengue vector control: cluster randomised trial. *British Medical Journal* **338**: 1477-86.

Walker, K. and Lynch, M. (2007) Contributions of Anopheles larval control to malaria suppression in tropical Africa: review of achievements and potential. *Medical and Veterinary Entomology* **21**: 2-21.

Walsh, J. F. (1983) Control of simuliid black flies. In: Youdeowei A. and Service M. W. *Pest and Vector Management in the Tropics*. Cambridge University Press, Cambridge. pp. 280-7.

WHO (2005) 'Guidelines for laboratory and field testing of mosquito larvicides.' WHO/CDS/WHOPES/GCDPP/2005.13.

WHO (2010) 'Generic risk assessment model for insecticides used for larviciding.' WHO/HTM/NTD/WHOPES/2010.4 (web only).

Yap, H. H., Lee, Y. W. and Zairi, J. (2002) Indoor thermal fogging against vector mosquitoes with two *Bacillus thuriengiensis israelensis* formulations, Vectobac ABG 6511 water dispersible granules and Vectobac 12AS liquid. *Journal of the American Mosquito Control Association* **18**: 52–6.

Yaméogo, L., Tapsoba, J. M. and Calamari, D. (1991) Laboratory toxicity of potential blackfly larvicides on some African fish species in the Onchocerciasis Control Programme Area. *Ecotoxicology and Environemental Safety* **21**: 248–56.

Yaméogo, L., Elouard, J. M. and Simier, M. (1992) Typologie of susceptibilities of aquatic insect larvae to different larvicides in a tropical environment. *Chemosphere* **24**: 2009–20.

Yaméogo, L., Crosa, G., Samman, J., et al. (2001) Long-term assessment of insecticide treatments in West Africa: aquatic entomofauna. *Chemosphere* **44**: 1759–73.

Yaméogo, L., Resh, V. H. and Molyneux, D. H. (2004) Control of river blindness in West Africa: case history of biodiversity in a disease control program. *EcoHealth* **1**: 172–83.

Chapter 6

Integrated Vector Management

There has been acceptance that a single approach to reducing vector borne disease inevitably fails. The prescribing of a medicine works well until the parasite has developed resistance to the drug, and the application of insecticides to control the vector likewise leads to resistance of the vector to the insecticide. There is thus a need to integrate different control strategies in a harmonious manner, so that the level of the vector borne disease is at a low level. Public health needs a similar strategy to Integrated Pest Management (IPM), developed in agriculture, when it was realised that over reliance on pesticides led to selection of resistant pests and adverse effects on the environment.

The World Health Organisation has developed a Global Strategic Framework for Integrated Vector Management (IVM) (WHO, 2004), sometimes referred to as integrated disease management (IDM), which aims to strengthen vector control in a manner that is compatible with national health systems. IVM is defined as a rational decision-making process for the optimum use of resources for vector control (Beier et al., 2008). The aim is to utilise evidence-based decision-making to rationalise the use of human and financial resources and organisations, both government and non-government, and village communities to ensure ecologically sound and sustainable control of vector borne diseases. In some countries, IVM is associated only with one disease such as malaria, but the aim is to integrate action against the range of diseases that can affect a community. Thus, in parts of West Africa, it is being increasingly recognised that alongside controlling Anopheline mosquitoes, it is essential to control black flies (*Simulium damnosum* complex) if people are to be able to work and improve productivity of their area. The diseases and vectors that need to be considered in an IVM programme will depend on local situations. Similarly the tools used in an IVM programme will need to be selected on the basis of local knowledge of the vectors, especially their biology in terms of biting behaviour and larval breeding sites.

In many countries, development programmes, including irrigation schemes for agriculture, construction of hydro-electric dams and other construction

Integrated Vector Management: Controlling Vectors of Malaria and Other Insect Vector Borne Diseases, First Edition. Graham Matthews.

Table 6.1 Components of an Integrated Vector Management programme for combating malaria.

Cultural control	Biological control	Chemical control	Disease treatment
Drainage of larval breeding sites	Use of larvivorous fish	Indoor residual spraying	Quinine
Screening of houses	Bio-pesticides (*Bacillus thuringiensis, B. sphericus*)	Treatment of bed nets	Chloroquin (no longer recommended)
Untreated bed nets		Space treatment (thermal or cold fog)	Artemesin Combination Therapy (ACT)
Polystyrene beads		Barrier sprays Larvicide application (granule or spray) Impregnated clothing, bed nets and curtains Repellents	Prophylactics

activities, provide additional ecosystems in which vector species can flourish. Thus, in developing its strategic plan, the WHO has recognised that IVM needs the input of organisations, other than Ministries of Health, to ensure that area-wide implementation of vector control is achieved. WHO has also recognised that a key feature of successful control programmes has been effective management with robust systems for monitoring, evaluation and reporting, so that when problems arise, there are procedures to identify needs and correct those problems. Key components of an IVM programme for malaria are shown in Table 6.1. Some components of this programme are also relevant to other vector control programmes. WHO has prepared a handbook on IVM (WHO, 2011) aimed at managers of vector borne disease control programmes in planning and implementing IVM.

How the control techniques fit in with other aspects of an IVM programme are shown in Figure 6.1. The choice of which components are needed does depend on good data on the local species, their biology and their susceptibility to insecticides. The primary aim must be to break the disease cycle, but having achieved that, it is also important that the vectors continue to be controlled where their presence in large numbers creates a major nuisance affecting man's activities at work and leisure. Vector control also needs to be maintained when the proportion of a human population affected by a disease has been significantly lowered, otherwise resurgence of the problem will occur.

Before the advent of modern insecticides, the main control measures were cultural and involved drainage around houses and screening windows and doorways to reduce access of mosquitoes into houses. Following the realisation that quinine was an effective cure of malaria, much attention was given to

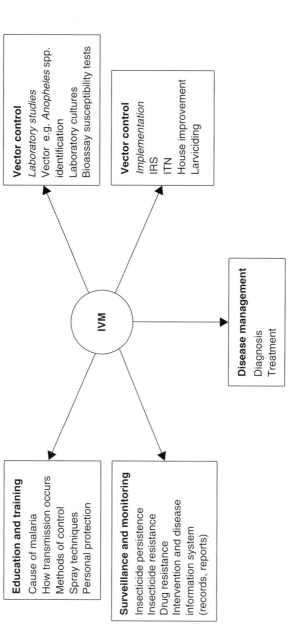

Figure 6.1 Diagram with components of Integrated Vector Management.

research to find a cheap alternative drug, leading to the development of chloroquine, which was then used extensively until the malarial parasite became resistance to it.

From the 1950s, it was hoped to eliminate malaria by spraying the inside of houses with DDT. There were many early successes with this approach, where housing was improved and mosquitoes were not active all year due to cold or dry winters. In the tropics, there was success in some areas, such as Sri Lanka, where the incidence of malaria decreased very significantly, but vector control was not sustained and cases of malaria again increased. Generally poor housing and mosquito activity over longer periods, as well as selection of mosquitoes resistant to DDT and inadequate funding, led to the abandoning of the eradication strategy. WHO then emphasised case detection and treatment rather than house spraying, but this resulted in an increase in malaria. Roberts (2010) illustrated the decline in house spraying with re-emergence of malaria in South America (Figure 2.1).

More recently, the emphasis has been on using insecticide treated bed nets (ITN). This has been successful in reducing mortality of young children, and in a study in The Gambia, Alonso et al. (1991) reported that following intervention with ITNs, the overall mortality and mortality attributable to malaria of children aged 1-4 in the intervention villages was 37 and 30%, respectively, of that in the non-public health care (PHC) villages with no bed net intervention.

However, ITNs need to be used by all members of a community to have an overall impact on malaria. Even then their effectiveness is limited, as most people are not under the net throughout the period when the vector is active. This has led to greater interest in using indoor residual spraying (IRS), but there has been reluctance by health authorities to implement IRS programmes. Unlike prescribing a net, adoption of the procedures that were used with DDT requires an extensive spray programme, but medical staff have not been trained in this subject. This illustrates the need, within the WHO IVM strategy, to involve other agencies to implement aspects of the vector control programme. Rather than relying on one control technique, integration of several different control techniques will require different skills, if an IVM programme is to be successful.

A large village-scale trial (Matthews et al., 2009) suggested that there is an advantage in using both IRS and ITNs, although mathematical modelling has shown that when using DDT as an IRS treatment in all houses and ITNs are not used, transmission can be reduced up to 10-fold more than the reduction achieved with ITNs alone (Yakob, 2010). It is pointed out that where both are used and there is a strong repellent effect, for example, when there is a DDT deposit on walls, the likelihood of mosquitoes touching an ITN is reduced and conversely an ITN reduces the likelihood of a blood-fed mosquito resting on a sprayed wall. To combine the two control techniques suggests that in search-ing for new insecticides, a chemical which is not repellent would be preferable to using a repellent insecticide, such as DDT.

Unfortunately very few people in poor countries, where vector disease control is most needed, have the skills needed to both manage and implement vector control. This is why there has not been sufficient coverage of control programmes in many countries to have a major impact on vector borne diseases. Hitherto, most control programmes have been initiated by governments ('vertical planning'), rather than involving local communities in the developing a sustainable strategy (Service, 1993). The adoption of different control techniques and targeting where they are applied in rural and urban areas requires individual countries to create national strategic plans that will prioritise education and training at both central organisations and in districts and villages, to ensure gaps in manpower will be adequately met and the vector control programme sustained.

Chemical control techniques need specific skills to implement and monitor the vector control programme. Unfortunately, the same insecticides have inevitably been used in both public health and agriculture, thus human disease vectors are exposed to selection for resistance to the various modes of action outside the public health sector. Ideally, certain insecticides should be restricted for use in public health, but in the global economy, the cost of development of new insecticides necessitates their widespread use with registration on major crops. Only then, if the insecticide is established, will it be considered for the relatively small public health market. Initiatives funded by the Gates Foundation have sought ways of introducing new insecticides, but this has so far relied on inventories of major agrochemical companies and improved formulation of some insecticides to allow more effective use in vector control.

The lack of a range of new insecticides of low mammalian toxicity necessitates greater emphasis on alternatives to chemical control, where this is feasible. Future research (Chapter 8) may enable new technologies to be introduced, but in the meantime more emphasis is needed on educating people to be aware of the importance of cultural controls.

Cultural controls

House design

The traditional house in the tropics has been designed to enable ventilation at night and tends to leave a large space between the top of the wall and the roof. This gap, at the eaves and usually many other openings in the walls, other than windows and doorways, allow mosquitoes to have free access to the houses. Poor quality walls of mud houses should be easily repaired by filling the holes with fresh mud, while the gap between the wall and roof could be covered with a screen. Early studies had shown that anopheline mosquitoes readily enter houses through the eaves and this has been confirmed in recent studies in Tanzania (Ogoma et al., 2010).

Schofield and White (1984) and Schofield et al. (1990) discussed the importance of house design and advocated the use of a ceiling in rooms to reduce mosquito entry. In a trial in The Gambia, Lindsay et al. (2003) confirmed that house entry by *Anopheles gambiae* was reduced to 59% by the presence of a plywood ceiling, 79% by a synthetic-netting ceiling, 78% with the netting treated with insecticide (deltamethrin), and 80% by using a plastic insect screen. Merely closing the eaves with mud was less effective, although in a village in Cameroon, fixing netting across eaves (Figure 6.2a) combined with a barrier mist spray on vegetation around the houses (Figure 6.3) did reduce

(a)

(b)

(c)

Figure 6.2 (a) Improving houses, by screening the gap between the wall and the roof; (b) modern house with louvre windows, but no screen to prevent insect entry; (c) verandah of house protected by a screen; (d) and (e) screen at door (photos: Steve Lindsay).

(d)

(e)

Figure 6.2 (*Continued*)

Figure 6.3 Barrier treatment of vegetation around houses (photo: Didier Baleguel).

mosquitoes entering the houses (Matthews et al., 2009). The netting was not treated with insecticide, but as shown in Chapter 4, a mesh fabric developed as a durable wall lining could be very effective as a screen across the eaves, through which anopheline mosquitoes enter houses (Njie et al., 2009). Some new houses are fitted with louvre windows, but in the example shown in Figure 6.2b, there is no screen, although the window is protected with bars. A screen should be fitted to any verandah on a house (Figure 6.2c) and doors should be protected by a secondary screen door (Figures 6.2d, e), which needs to be kept closed, even if the main door is left open.

The main reason why few villagers do not put ceilings in their houses or close up the eaves is due to reduced ventilation and the extra cost involved, although surveys in The Gambia have shown that they perceived that ceilings improved the functionality and beauty of their houses and there was less disturbance caused by mosquitoes.

Ogoma et al. (2009) reported that a survey in Zanzibar showed that over 80% of houses had screens, recognising their importance of keeping mosquitoes out of their houses. Other studies have shown a reduction in anaemia with improved houses (Kirby et al., 2009). Similarly the transmission of arboviruses by mosquitoes can be reduced when housing is improved.

In South America, as the vector of Chagas disease can hide in cracks and crevices, a key requirement is to improve the quality of the wall surface of houses to remove places in which the Triatomine bugs can hide (Garcia-Zapata, 1992), the cost of which is mostly for labour, unless the work is done by the house owner. In addition to house improvement, the buildings used for keeping animals also need to considered and sited some distance from where humans are living (Arata et al., 1994).

Although the need for improved housing might be considered essential, the perception of some of the poorer people is that they lack the motivation or means to improve their own housing. This emphasises the need for income generation within a locality, so that once poverty is alleviated they will want better houses. Thus within IVM, chemical control initially funded from outside to reduce vectors and raise health care is needed, but needs to be followed up with subsequent encouragement to improve housing to sustain the impact of vector control.

Drainage and water management schemes

Prior to the use of DDT in the 1940s, one of the main methods of reducing vector borne disease was to drain water from around houses. Large areas of wet land were drained in some countries to reduce mosquito populations or management practices were implemented to reduce breeding of mosquitoes. Malaria was endemic in the Tennessee Valley, USA, when in the 1933 the Tennessee Valley Authority (TVA) was established to provide navigation, flood control, electricity generation, fertiliser manufacturing and economic development in the Tennessee Valley, a region particularly impacted by the Great Depression. A major factor in reducing mosquitoes was achieved by controlling the water flow, and by clear-cutting the margins of bodies of water to reduce or eliminate mosquito habitats. Some applications were made with oils and inorganic larvicides, but mosquito control in the TVA area was not brought about by insecticides. New agricultural methods were introduced into traditional farming communities, and houses were improved with mosquito screens. When DDT was available, it was applied aerially in 1943 at 0.056-0.0112 kg/ha as a larvicide, replacing Paris Green dust completely by 1946, and boat applications of kerosene and black oil mixtures were discontinued in 1949. Initially, fixed wing aircraft were used, but these were subsequently replaced by helicopters also used for transmission line patrol, with DDT replaced by temephos in 1967. DDT was also applied inside houses, but with the subsidence and virtual disappearance of malaria by 1950, supplementary malaria control was discontinued (Derryberry and Gartrell, 1952; Gartrell et al., 1981). The aim of the programme was to suppress mosquito larval populations at specific sites within the TVA area, so that mosquito populations are below the level that enables disease transmission and human annoyance to occur (Kitron and Spielman, 1989).

The development within the Tennessee Valley illustrates how co-ordination was achieved by setting up an authority with overall responsibility throughout the area, integrating the different organisations that benefited from the development. Drainage schemes, particularly in urban and peri-urban areas, can also reduce the risk of many water borne diseases such as diarrhoea (Esrey, 1996).

Similarly in Florida, drainage schemes were initiated as early as 1919, and especially from 1933-41, when more than 2,400 km of drainage ditches were dug for mosquito control. Subsequently from 1945-49, IRS with DDT was carried out and since then mosquito control programmes have emphasised larviciding in relation to the rainfall pattern and adulticiding with space treatments in response to high numbers of mosquitoes. The effective control of mosquitoes since 1958 (Figure 6.4) has enabled Florida to build up a major tourist industry.

The TVA programme has not led to many environmental issues (Gartrell et al., 1972), but there have been protests in some countries by those concerned about the impact of loss of marshes and other natural habitats of bird populations. In the USA, water management programmes are designed to conserve water and stock aquifers as well as preserve wild-life habitats. As pointed out by Kitron (1987), anti-malaria campaigns need to adopt flexible strategies incorporated into local health services and inter-related with agricultural practices.

WHO published guidelines on methods of environmental management for mosquito control (WHO, 1982), which are still applicable today. In an assessment of environmental management for malaria control in Asia, a distinction was made between environmental modification and environmental manipulation (Lindsay et al., 2004). Modification requires significant investment for

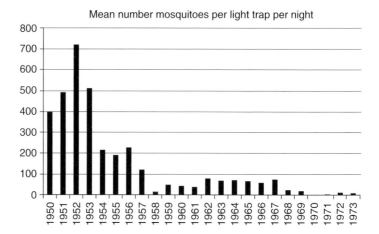

Figure 6.4 Historical data redrawn from *Florida Public health Pest Control Manual*, showing the decline in mosquito numbers as the State organised mosquito control areas. Tourism increased from 1958 onwards, to become a major part of the State's economy. Light traps were of an older design and were not baited with dry ice. The data is thought to refer to urban areas and not nearby wetlands.

permanent changes such as landscaping, drainage, land reclamation and filling in land liable to collect significant amounts of water. In contrast, manipulation is a recurrent activity, requiring proper planning and operation, such as removing aquatic weeds from irrigation and drainage canals, and may be part of good agricultural practice. However, some infrastructural development may be required. Nevertheless, much can be done to reduce areas that collect water in the vicinity of houses, by using a shovel to fill in the area with soil. As mentioned in the previous chapter, larviciding can be integrated in the programme where ponds or other areas of water are important in the local environment. Overall environmental management for vector control is not a replacement for other control strategies, but is complementary to chemical control, by reducing the need for too many spray applications.

Personal protection

Impregnated clothing

Individuals should always take responsibility for protecting themselves as much as possible, thus wearing clothing covering as much of the body reduces the area of skin exposed to vectors. Impregnation of clothing with insecticides clearly has a role for the military when on active service, but can be valuable for others when in areas where control operations are not feasible or are not permitted to protect natural habitats.

In situations where large insect populations are a nuisance, wearing insecticide impregnated clothing may have a better effect on reducing the nuisance than just covering the skin with a repellent.

Insecticide treated bed nets

The use of ITNs (Chapter 4) has become a major part of programmes to reduce the mortality of young children due to malaria, although there has been some concern that if a child is not exposed to malaria, it will not build up any immunity to the disease. As the use of treated bed nets is adopted more widely, the effect on mosquito populations is more significant. Nevertheless, it is only protective when the persons are under the net, so needs to be used as only one component of IVM. When houses are improved with screens, the use of a net is more optional, although in areas of endemic malaria, their use remains important.

Repellents

Repellents will vary from those that truly repel mosquitoes from entering a place, those that are really deterrents, so that when the mosquito lands on a treated surface, it is deterred from staying and thirdly, those that on landing, however briefly, pick up a lethal dose of chemical. Use of repellents to reduce the biting of mosquitoes dates back a long time, with some of the old techniques still being used. Mosquito coils are made with natural pyrethrum integrated in an extruded ribbon of wood dust, starch and various additives. When ignited, the coil will burn slowly over several hours and release a smoke that deters or repels a high proportion of mosquitoes from entering treated huts (Figure 3.2). This reduces biting and alleviates the nuisance of mosquitoes during the night (Smith et al., 1972).

The use of repellents is usually associated with individuals putting a cream containing DEET or other repellent on their skin to stop mosquitoes biting. The repellent DEET (*N,N*-diethyl-3-methylbenzamide) has been widely used to protect from mosquito bites. Unfortunately its effect does not last long (<3-6 h) and some people find it irritates their skin. In cage studies, the mean duration of protection (in hours) from bites (MDPB) of *Aedes aegypti* and *Anopheles quadrimaculatus*, using DEET on a 50 cm^2 area of healthy human skin, showed that for both species, the biting rate was inversely proportional to mosquito density and the MDPB. The least protection from biting was in large cages with a high density of mosquitoes, while protection was better with few mosquitoes in the medium-sized cage used in these tests (Barnard et al., 1998). Alternatives to DEET were not considered to be as effective (Fradin and Day, 2002). Barnard (1999) found that none of the essential oils he tested prevented mosquito biting, when used at 5-10% concentration, although thyme and clove oils were the most effective. However, their acceptability may be limited due to their odour and as they can cause irritation of the skin.

Other tests have shown that DEET and certain other repellents can give protection from some species for several hours. In Bolivia, Moore et al. (2002) demonstrated that better protection (96%) than DEET was obtained for 4 hours with a repellent based on eucalyptus containing 30% para-menthane-3, 8-diol (Menthoglycol). Such protection integrated with use of bed nets as the malaria vector *Anopheles darlingi* bites during the evening. In one test in Australia, two formulations of DEET were better than picardin, but all gave good protection against *Culex annulirostris* (Frances et al., 2002). Repellents are also available as aerosol sprays and one report confirmed that all the products tested did cause knock-down and significant mortality in 24 hours, with the synthetic organic repellents having a more rapid effect than botanical products (Xue et al., 2003). WHO has issued guidelines for efficacy testing of mosquito repellents applied on human skin (WHO, 2009).

Recent studies have shown repellence can be increased by mixing two compounds, thus a mixture of 6-methyl-5-hepten-2-one and geranylacetone in a 1:1 ratio presented at low concentrations exceeded the repellence of DEET over several hours (Logan et al., 2010). The behavioural response of the mosquitoes was significantly affected by the ratio of these compounds, indicating that mosquitoes can detect and respond to differences in natural repellent chemicals emitted from hosts. A suitable formulation of this mixture is a potential new repellent for personal protection.

When a repellent effectively protects an individual, does the mosquito move to the next person to bite? In an experiment, one or the other or both individuals in pairs, 1 metre apart, had no repellent (mineral oil in ethanol was used as a control) and this showed that more mosquitoes landed on the untreated person, when the partner had the repellent (di-ethyl toluamide –DEET) (Moore et al., 2007). Thus, within a locality, all would need to use a repellent to reduce the risk of malaria, although the degree of protection may be affected by the extent of the choice between those using repellents and non-users.

However, a new insecticide, metofluthrin, which vaporizes at room temperature, is highly effective against mosquitoes and when used on paper substrates as 'emanators', its use has been shown to reduce landing rates by more than 85% compared to untreated controls (Lucas et al., 2007).

The use of repellents for protecting farmers and fishermen exposed to large populations of black flies has not been evaluated, but development of a suitable emanator would significantly help combat the continual nuisance of biting flies while working.

However, there are other possible ways in which a repellent might be used in an IVM programme. Could a repellent applied to a water container stop or reduce oviposition? Piperidine compounds, compared to DEET, have shown that fewer mosquito eggs were laid under experimental conditions in which 1-litre black plastic containers were treated with different concentrations of repellent. Using 0.1% concentration, there was more than 50% deterrence for up to 21 days against field population of *Aedes albopictus* in Florida (Xue et al., 2001).

Another approach might be to treat bed nets with a repellent so that combined with IRS the net repels the mosquito from the bed net to a treated wall surface.

Barrier treatments

Where there is improved housing or people decline to have indoor spray treatments, the numbers of mosquitoes entering houses can be reduced by 'barrier' treatments, in which a residual insecticide is applied to vegetation in the vicinity of houses. Where barrier treatments have been applied, the high velocity air stream from motorised mist blowers (Figure 6.3) has been used to project spray into trees and also enable a larger area to be treated quickly. They have also been used to treat the outside of houses, especially to spray up into the eaves of a building.

Barrier treatments have also been evaluated on small plots in a desert area with sparse vegetation (Britch et al., 2009), when numbers of mosquitoes caught in treated plots was 80% lower than in control plots. Bioassays on samples of vegetation indicated residual activity extending over 28 days. Apart from the effect on mosquitoes, barrier treatments usually have a significant effect on flies generally.

Another type of barrier treatment has been the use of polystyrene beads in pit latrines to prevent larval survival. In Zanzibar it was shown that vector control, combined with mass treatment against Bancroftian filariasis with diethylcarbamazine (DEC), was more effective than relying only on the drug treatment (Maxwell et al., 1999).

A new approach to barrier treatments has been the use of an attractive toxic sugar bait (ATSB) prepared with fruit juices (guava, melon), honey, sugar and a stabiliser/preservative mixed with 1% (w/v) boric acid as the toxicant. The ASTB was applied with a knapsack sprayer at $80\,ml/1m^2$ spots at distances of 3m apart around ponds and rice paddies. Sampling with light traps showed a significant reduction of over 90% male and female *An. gambiae* following application of a toxic bait, compared with an area treated with bait without a toxin. By adding a colour dye, it was shown that over 50% of the trapped mosquitoes had fed on the baits (Muller et al., 2010).

Implementation of IVM

At present there are many people in government departments and non-governmental organisations advocating certain components of IVM, such as the distribution of ITNs, and having an impact within their own financial and logistic constraints. National programmes are often devised by government without using other resources or involving the private sector in a co-ordinated programme that will have an impact on productivity. A national IVM forum is needed to bring all interested sectors of society together and agree a programme (Figure 6.5). Within a country, there may need to be a separate authority for a particular region, thus in the USA, the TVA was a key to development of the area and control of vector borne disease. Similarly, regional cooperative action was taken by the North American Regional Action Plan (NARAP) on DDT to

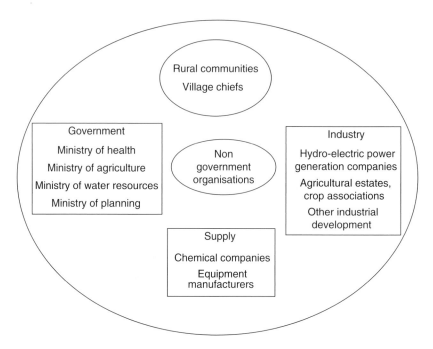

Figure 6.5 Organisations that need to be involved in an area-wide IVM programme.

achieve malaria control in Mexico, involving environmental management practices to eliminate mosquito breeding sites (Chanon et al., 2003).

The actual components used in IVM need to be based on local vector ecology, epidemiology and resources, and this also necessitates operational research on a sufficiently large scale to validate the suitability of any individual control technique. Thus, in adopting chemical control, tests on the susceptibility of the vector to the insecticides and monitoring the impact in the field provide the basis for further extension at a district, regional and eventually national programme.

All this depends on a having sufficient trained staff, so a programme of training will be a key step to ensure that there are adequate human resources to implement and sustain an IVM programme. Capacity-building also requires providing adequate laboratory facilities and financial support above the actual cost on chemicals, bed nets and other components used in the programme, to ensure the programme is technically supported. A key problem in many countries is that there are not sufficient people with appropriate training to regulate and control the use of pesticides throughout the life-cycle of the pesticide from purchase to safe disposal of materials after their use.

National, and in some cases regional, support is needed in terms of legislation and regulation, to ensure that materials can be transported readily and safely at the appropriate time to where they are needed. Where insecticides are used, the appropriate formulations and equipment need to have been registered and supplied in the right quantities and quality on time, to allow distribution down to districts before the optimum time for control operations.

Figure 6.6   Clearing a drainage channel in Zambia (photo: Peter Mukuka).

An example of IVM at Copper mines in Zambia
(data from Peter Mukuka)

Actions to reduce the number of people affected by malaria by adopting more than one vector control technique is well illustrated by the programme at Mopani copper mines in Zambia, which runs Nkana and Mufulira Mine sites. Mopani Copper Mines took over the running of two mine sites in April 2000 from Zambia Consolidated Copper Mines, Ltd (see also RBM, 2011).

In 2000, the mine management decided to take action against the large number of cases of malaria within the mining community. The community was living in brick houses with corrugated iron or asbestos roofs, but without gaps in the eaves to allow access of mosquitoes (Figure 6.7). At one mine site, Mufulira, the initial effort was concentrated on environmental management, by making water ways (Figure 6.6) to drain surrounding *dambos* (wet areas) to reduce mosquito breeding, clearing township drains, and then larviciding ponds around the mine. This was the main control technique from 2000–03, before commencement of IRS and has been continued. Temephos was sprayed as the larvicide. However, since 2004, IRS has been carried out annually, using either λ-cyhalothrin as a 10% wettable powder (WP) formulation supplied in 62.5 g sachets or sometimes deltamethrin, aimed at treating at least 80% and above of the houses. During this period, thermal fogging was carried out occasionally each year on a small scale in selected areas, including some houses, culverts, under bridges and in some dense vegetation such as hedges. ITNs were not distributed by the company. In addition, public education on malaria is conducted at all Health Centres and at other fora. The company also produces education material on malaria, in order to increase public awareness about the disease and what they can do to protect themselves.

(a)

(b)

Figure 6.7 (a) and (b) Houses for mining staff at Mopani Mines, Zambia (photos: Peter Mukuka).

There was no mosquito control programme in areas adjacent to the mine until recently, when IRS was introduced with DDT or pyrethroid insecticides being used (see Chanda et al., 2008 on the introduction and expansion of IVM in Zambia, and Chizema-Kawesha et al., 2010 for review of Zambia programme). Thus, when employees and their dependants went on leave or were away from the mine, repellents were supplied to protect exposed skin. Repellents were also used by employees if working on a night shift or were prone to malaria attacks.

The decline in cases of malaria, since vector control interventions were used, shows how effective the programme has been. The effect of just larviciding from 2000–03 at one mine is clearly shown (Figure 6.8). The

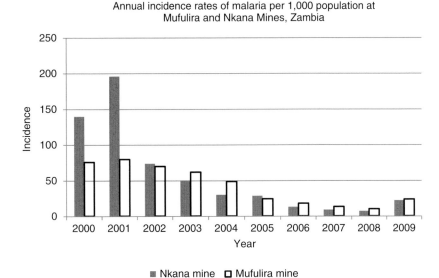

Annual incidence rates of malaria per 1,000 population at
Mufulira and Nkana Mines, Zambia

■ Nkana mine   □ Mufulira mine

Figure 6.8 Annual incidence rates of malaria at Mufulira and Nkana mines,
Zambia in relation to the periods with larviciding and IRS (data: Peter Mukuka).

increase in the number of cases in 2009 was due to less financial support for
vector control following the global credit crisis problems. All cases of malaria,
whether at the out-patients department or admitted to the hospital, have
been recorded and confirmed by blood smear tests, although the rapid
diagnostic test is now available. All positive admitted cases are investigated to
determine the source to check on the status of the house and whether it had
been sprayed. Most cases of malaria occur during the wet season (Figure 6.9).
In carrying out all these interventions, the company works closely with the
Ministry of Health through the National Malaria Control Program and District
Health Management Teams.

This integrated vector control programme has been successful, due to:

- Drainage works reduced significantly the extent of breeding sites within
  the vicinity of the mine.
- Larviciding reduced mosquito larvae survival in areas that could not be
  drained.
- House design limited access for mosquitoes.
- IRS controlled mosquitoes that did enter houses.
- Space treatments reduced mosquitoes at selected sites and reduced
  exophilic biting.
- Repellents gave some protection of those who had to be out at night or
  leave the mine site.
- The number of people that needed treatment was significantly reduced.

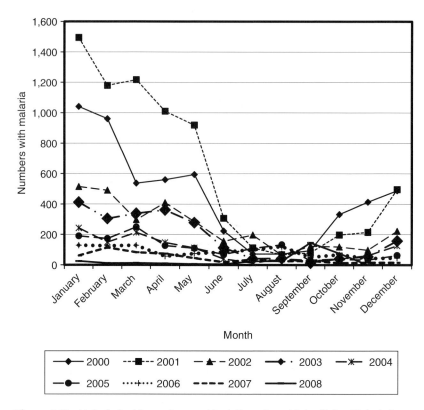

Figure 6.9 Malaria incidence by month at the mines (data: Peter Mukuka).

It also showed that when financial constraints reduced the programme, that cases of malaria soon increased; emphasising the importance of a sustained approach over a large area to reduce the re-invasion of mosquitoes with the malarial parasite.

Success of an IVM programme will be shown by a significant decline in the number of people with a vector borne disease. This will depend on having trained staff to take a blood sample on a microscope slide and know how to examine it to detect the malarial parasite. There are now rapid diagnostic kits increasingly available, so that an accurate assessment is made of the incidence of malaria. Often in the past sick people with a fever might have been recorded as a malaria patient without a proper diagnosis. With malaria, it is crucial that any person with a fever is accurately diagnosed if they have malaria so that they get the correct treatment (Figure 6.10). Inaccurate diagnosis results in inappropriate and overuse of the drugs, such as artemisinin combination therapy (ACT). Once vector populations are under control, surveillance needs to be sustained (Figure 6.11) so that if vector populations rise or a vector invades the area, action can be taken quickly. Similarly, surveillance of the human population is needed to check infection rates, especially if there is movement of people from areas where malaria is still endemic.

Figure 6.10 Mother with sick child at a doctor's clinic.

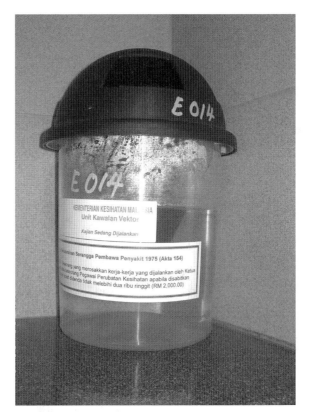

Figure 6.11 Trap at Malaysian airport for surveillance of mosquitoes (photo: Chung gait Fee).

Table 6.2 Different ways of expressing costs involved in vector control.

Cost usually expressed in $	Numerical changes
per person protected per year	disability adjusted life year (DALY) prevented (morbidity and mortality combined)
per household treated per year	years of life lost (YLL)
per case prevented	years lived with disability (YLD)
per death prevented	vector population reduction
per year of additional life	reduction in malaria incidence
per area (e.g. standard of 10 km²)	changes in sporozoite rates, biting rates, persons with anaemia

Costs

The financial cost of controlling vectors and impact of control on the diseases can be expressed in a number of different ways (Table 6.2), so comparison between the different control techniques depends on similar data being available. From a medical point of view, disability adjusted life year (DALY) is one of the most widely used measurements.

The aim of vector control programmes is initially to reduce the population of vectors, so the cost to treat an area is perhaps the most relevant, as the cost in terms of number of cases of malaria or death averted will depend on the number of people within an area, the population of mosquitoes and the incidence of malaria within that population. Costs for some vector control programmes will also depend on the seasonality of the vector, so in areas with a distinct period over which protection will be required, a programme will vary from a few months each year to the whole year to depending on rainfall patterns.

The Roll Back Malaria programme (RBM) has issued a calculation programme in Excel – http://www.rollbackmalaria.org/toolbox/tool_malaria/Costing.html. This has sections relating to drug treatments, bed nets and using insecticides as larvicides and for IRS treatments, but the costings refer principally to the capital cost of the equipment and consumable costs and ignore many of the other factors that need to be considered (Table 6.3).

Costs per person protected, unless otherwise indicated, which have been extracted from some published documents (Table 6.4), are not directly comparable, as they are derived in different ways and do not include all the same variables.

Yukich et al. (2008) illustrates the variation in costs between countries using the average annual economic cost for ITN and IRS programmes (Tables 6.5 and 6.6). One of their main conclusions was that the use of long-lasting insecticidal nets (LLIN) with 3 or 5 years of useful life is more cost-effective than the use of conventional nets and much more cost-effective than the application of IRS.

Guyatt et al. (2002) stated that the average retail price for a family or double-size bed net bought from the open market in Africa was between US$5.00 and

Table 6.3 Factors in costing different interventions.

IRS	ITN	Both IRS and ITN
Cost of sprayers at country of use	Cost of net (re-treatable or LLITN)	Operational research, including bioassays
Cost of persons doing the spraying (salaries and per diem costs)	Cost of treating an untreated net or re-treatment (Insecticide + ancillary items)	Education – sensitise users about the control treatments
Ancillary equipment (PPE etc.)	Cost of ensuring net is used (supply of nails, string to hang the net)	Management/ co-ordination
Maintenance of equipment (spare parts and workshop staff costs)	Transportation to distribute to villages	Storage of equipment, nets, insecticides
Transport costs to villages (getting equipment and personnel to individual villages)	Training users	Monitoring and evaluation of impact
Training spraying staff		Quality assurance
Medical check on spray operators		Recycling and disposal of items Importation costs, taxes Environmental impact assessment

Table 6.4 Some costs given in published documents.

Reference	IRS	ITN
Unpublished (Cameroon)	3300 fcfa ($7.16)[#]	1442 fcfa ($3.13)[##]
Guyatt et al., 2002	$0.86	$4.21 ($2.34*)
Guyatt et al., 2002		$4–5 (net)** + $1–2 (treatment)
Curtis and Maxwell, 2002		$1.03[x]
Guyatt et al., 2002 [2]		$8.42 (NGO programme) – $30 (including delivery costs)
Bhaia et al., 2004	$ 87[xx]	$ 52[xx] (in India)
Graves, 1998		$15.75 (per case averted)[xxx]
Yukich et al., 2008	$3933–4357	$998–2926 (per death averted)

[#] Cost of insecticide per house per treatment.
[##] Assumes net lasts 3 years and is treated twice a year. Screening houses and outdoor barrier treatment with insecticide was approximately $30 per house.
* Costs of nets sensitive to number sold in a community.
** Price for untreated nets in bulk when delivered.
[x] Nets expected to last 4 years with annual re-treatment (Guyatt and Snow[2] argue that these costs are exclusive of true delivery costs).
[xx] Per malaria case averted.
[xxx] This was compared with $3.71 for using an effective vaccine.

Table 6.5 Cost of ITN programmes.

Country	Average cost per ITN distributed	Average cost per TNY**	Cost per death averted (ITN)	Cost per DALY averted (ITN)
Eritrea	4.74	1.43	1,722	52
Malawi	3.36	3.04	1,222	37
Tanzania	4.80	2.17	1,745	53
Senegal	8.05	6.05	2,926	89
Togo*	3.23	3.23	1,174	36

* First value assumes an average net cost of USD 4.33, while the second value assumes a cost of USD3, which is more representative for a conventional net.
** = Treated Net Year.

Table 6.6 Cost of IRS programmes.

Country	Cost per person protected (whole population)	Cost per under-five child protected	Cost per death averted	Cost per DALY averted
KwaZulu-Natal	3.27	23.96	4,357	132
LSDI/MOZAL	3.90	21.63	3,933	119

US$10.00, but if nets were bought in bulk, then prices could be reduced to US$4.00–5.00. The cost for the insecticide will also depend on the size of the net and whether the treatment is provided as part of a bulk purchase of insecticide or as a pre-packaged single treatment. Most of the cost data available in the literature is for permethrin, which is currently recommended to be administered every 6 months in areas that are not subjected to acute seasonal malaria risk. The cost of permethrin for a family or double net is around US$0.50–1.00 per treatment, i.e. US$1.00–2.00 per year.

In considering long-term programmes, Utzinger et al. (2001, 2002) estimated that in Zambia an integrated programme with strong emphasis on environmental management, rapid diagnosis and treatment of malaria and use of bed nets sustained over 20 years (1930–50) averted an estimated 14,122 deaths, 517,284 malaria attacks and 942,347 work shift losses in the mining area of the Copperbelt. The sound investment of US$ 11,169,472 (in 1995 US$) helped copper production revenues to increase and amounted to US$ 7.1 billion (in 1995 US$).

The costs mentioned in scientific publications, however, do not always reflect the reality of the situation in many countries, where the materials for vector control may be subject to import tax, port dues or other levies, in order to get them through Customs to where they are being used.

Development of new technology

In Chapter 8, developments of new technology aimed at vector control are discussed, but for any of these to succeed, it is important that they are effective and fit into an overall strategy. As Kilama (2009) has pointed out, this will require intensive research on the efficacy, bio-safety and risk/benefit analysis of candidate vector control tools, in comparison with existing tools, to establish by means of field studies their cost-effectiveness, acceptability and efficiency.

To achieve this will require strengthening of institutions in countries with malaria and other tropical diseases. At present there are few institutions in Africa with adequately trained staff to carry out extensive field trials. Kilama (2009) stated that trials of new vector control tools should also be subject to a bio-safety and ethical review. This would need to be carried out by an ethics review committee, which should include or co-opt medical entomology expertise, and follow guidelines that are needed to take a holistic approach. Furthermore, the researchers in vector control trials should undergo bio-ethics and bio-safety training. The outcome of such research will also require careful dissemination, so that successful vector control interventions can be made available in other appropriate areas with the disease.

In an attempt to assist in assessing the success of interventions, models have been developed, although those of necessity have to make various assumptions. In one model, Griffin et al. (2010) concluded that interventions using current tools, and in particular LLINs, can result in major reductions in *Plasmodium falciparum* malaria transmission and the associated disease burden in Africa, especially in low- to moderate-transmission settings, when vectors are primarily endophilic (indoor-resting), provided there is a comprehensive and sustained intervention programme. However, in high-transmission areas and where there are mainly exophilic vectors, other control techniques will be required. Ferguson et al. (2010) concluded that the transmission cycle of *P. falciparum* in the most intensely endemic areas cannot be broken by existing front-line vector control measures, such as insecticide-treated nets and residual sprays, and that greater investment is needed in understanding the ecology and evolution of the mosquito vectors that transmit malaria, with more emphasis on aspects of the mosquito life-cycle beyond the blood feeding processes that directly mediate malaria transmission.

Conclusion

While many would prefer a simple single control technique, the reality is that a multi-pronged approach is needed and this requires greater management skills to ensure that different interventions do integrate positively. The choice of different interventions will vary according to local circumstances. While much emphasis has been placed on sleeping under a net, greater efforts are needed to improve housing and drainage around dwellings and explore the potential of larviciding, and other control techniques as well as IRS.

Management needs to ensure staff are trained and take account of biological data (bioassays) on efficacy of pesticides and on all other aspects of pesticide use. Furthermore, it is vital that there is collaboration between the stakeholders and communication right down to village level, so that surveillance and control is effectively achieved within a wide area.

References

Alonso, P. L., Lindsay, S W., Armstrong, J. R. M., et al. (1991) The effect of insecticide treated bed nets on mortality of Gambian children. *Lancet* **337**: 1499-502.

Arata, A. A., Balderrama, F., Bermudez, H., et al. (1994) *Chagas in Bolivia. Project report: Community and Child Health Project (CCH)* Ministerio de Desarollo Humano, Secretarfa Nacional de Salud, Bolivia, US Agency for International Development, La Paz, P.L 480.

Barnard, D. R. (1999) Repellency of essential oils to mosquitoes (Diptera: Culicidae). *Journal of Medical Entomology* **36**: 625-9.

Barnard, D. R., Posey, K. H., Smith, D. and Schreck, C. E. (1998) Mosquito density, biting rate and cage size effects on repellent tests. *Medical and Veterinary Entomology* **12**: 39-45.

Beier, J. C., Keating, J., Githure, J. I., MacDonald, M. B., Impoinvil, D. E. and Novak, R. J. (2008) Integrated vector management for malaria control. *Malaria Journal* **7(suppl)**: 54.

Bhatia, M. R., Fox-Rushby, J. and Mills, A. (2004) Cost-effectiveness of malaria control interventions when malaria mortality is low: insecticide treated nets versus in house residual spraying in India. *Social Science and Medicine* **59**: 525-39.

Britch, S. C., Linthicum, K. J., Wynn, W. W., et al. (2009) Evaluation of barrier treatments on native vegetation in a Southern California desert habitat. *Journal of the American Mosquito Control Association* **25**: 184-93.

Chanda, E., Masaninga, F., Coleman, M., et al. (2008) Integrated vector management: the Zambian experience. *Malaria Journal* **7**: 164.

Chanon, K. E., Mendez-Galvan, J. F., Galindo-Jaramillo, J. M., Olguin-Bernal, H. and Borja-Aburto, V. H. (2003) Cooperative actions to achieve malaria control without the use of DDT. *International Journal of Hygiene and Environmental Health* **206**: 387-94.

Chizema-Kawesha, E., Miller, J. M., Steketee, R. W., et al. (2010) Scaling up malaria control in Zambia: progress and impact 2005-2008. *Journal of the American Mosquito Control Association* **83**: 480-8.

Curtis, C. F. and Maxwell, C. (2002) Free insecticide for nets is cost effective. *Trends in Parasitology* **18**: 204-5.

Derryberry, O. M. and Gartrell, F. E. (1952) Trends in Malaria Control Program of the Tennessee Valley Authority. *Journal of the American Mosquito Control Association* **1**: 500-7.

Esry, S. A. (1996) Water, waste and well-being: a multi-country study. *American Journal of Epidemiology* **143**: 608-23.

Ferguson, H. M., Dornhaus, A., Beeche, A., et al. (2010) Ecology: a prerequisite for malaria elimination and eradication. *PLoS Med* **7**: e1000303. doi:10.1371/journal.pmed.1000303

Fradin, M. S. and Day, J. F. (2002) Comparative efficacy of insect repellents against mosquito bites. *New England Journal of Medicine* **347**: 13-18.

Frances, S. P., Van Dung, N., Beebe, N. W. and Debboun, M. (2002) Field evaluation of repellent formulations against daytime and nighttime biting mosquitoes in a tropical rainforest in Northern Australia. *Journal of Medical Entomology* **39**: 541-4.

Garcia-Zapata, M. T. A., Marsden, P. D., Scares, V. A. and Castro, C. N. (1992) The effect of plastering in a house persistently infested with *Triatoma infestans* (Klug) 1934. *Tropical Medicine and Hygiene* **95**: 420-3.

Gartrell, F. E., Barnes, W. W. and Christopher, G. S. (1972) Environmental impact and mosquito control water resource management projects. *Mosquito News* **32**: 337-42.

Gartrell, F. E., Cooney, J. C., Chambers, G. P. and Brooks, R. H. (1981) TVA mosquito control 1934-1980 experience and current program trends and developments. *Mosquito News* **41**: 302-22.

Graves, P. M. (1998) Comparison of the cost-effectiveness of vaccine and insecticide impregnation of mosquito nets for the prevention of malaria. *Annals of Tropical Medicine and Parasitology* **92**: 309-410.

Griffin J. T., Hollingsworth, T. D., Okell L. C., et al. (2010) Reducing *Plasmodium falciparum* malaria transmission in Africa: a model-based evaluation of intervention strategies. *PLoS Med* **7**(8).

Guyatt, H. L., Kinnear, J., Burini, M. and Snow, R. W. (2002) A comparative cost analysis of insecticide-treated nets and indoor residual spraying in highland Kenya. *Health Policy and Planning* **17**: 144-53.

Kilama, W. L. (2009) Health research ethics in public health: trials and implementation of malaria mosquito control strategies. *Acta Tropica* **112S**: S37-47.

Kirby, M. J., Ameh, D., Bottomley, C., et al. (2009) Effect of two different house screening interventions on exposure to malaria vectors and on anaemia in children in the Gambia: a randomised controlled trial. *Lancet* **374**: 998-1009.

Kitron, U. (1987) Malaria, agriculture and development: lessons from past campaigns. *International Journal of Health Services* **17**: 295-326.

Kitron, U. and Spielman, A. (1989) Suppression of transmission of malaria through source reduction: Antianopheline measures applied in Israel, the United States and Italy. *Reviews of Infectious Diseases* **11**: 391-406.

Lindsay, S., Kirby, M., Baris, E. and Bos, R. (2004) 'Environmental management for malaria control in the East Asia and Pacific (EAP) region.' World Bank Health, Nutrition and Population Discussion Paper.

Lindsay, S. W., Jawara, M., Paine, K., Pinder, M., Walraven, G. E. L. and Emerson, P. M. (2003) Changes in house design reduce exposure to malaria mosquitoes. *Tropical Medicine and International Health* **8**: 512-17.

Logan, J. G., Stanczyk, N. M., Hassanali, A., et al. (2010) Arm-in-cage testing of natural human-derived mosquito repellents. *Malaria Journal* **9**: 239.

Lucas, J. R., Shono, Y., Iwasaki, T., Ishiwatari, T., Spero, N. and Benzon, G. (2007) US laboratory and field trials of metofluthrin (Sumione®) emanators for reducing mosquito biting outdoors. *Journal of the American Mosquito Control Association* **23**: 47-54.

Matthews, G. A., Dobson, H. M., Nkot, P. B., Wiles, T. L. and Birchmore, M. (2009) Preliminary examination of integrated vector management in a tropical rainforest area of Cameroon. *Transactions of the Royal Society of Tropical Medicine and Hygiene* **103**: 1098-104.

Maxwell, C. A., Mohammed, K., Kisumku, U. and Curtis, C. F. (1999) Can vector control play a useful supplementary role against bancroftian filariasis? *Bulletin of the World Health Organisation* **77**: 138-42.

Moore, S. J., Lenglet, A. and Hill, N. (2002) Field evaluation of three plant-based insect repellents against malaria vectors in Vaca Diez Province, the Bolivian Amazon. *Journal of the American Mosquito Control Association* **18**: 107–110.

Moore, S. J., Davies, C. R., Hill, N. and Cameron, M. M. (2007) Are mosquitoes diverted from repellent-using individuals to non-users? Results of a field study in Bolivia. *Tropical Medicine and International Health* **12**: 532–9.

Muller, G. C., Beier, J. C., Traore, S. F., et al. (2010) Successful field trial of attractant toxic sugar bait (ATSB) plant-spraying methods against malaria vectors in the *Anopheles gambiae* complex in Mali, West Africa. *Malaria Journal* **9**: 210.

Njie, M., Dilger, E., Lindsay, S. W. and Kirby, M. J. (2009) Importance of eaves to house entry by anopheline, but not culicine, mosquitoes. *Journal of Medical Entomology* **46**: 505-10.

Ogoma, S. B., Kannady, K., Sikulu, M., et al. (2009) Window screening, ceilings and closed eaves as sustainable ways to control malaria in Dar es Salaam, Tanzania. *Malaria Journal* **8**: 221.

Ogoma, S. B., Lweitoijera, D. W., Ngonyani, H., et al. (2010) Screening mosquito house entry points as a potential method for integrated control of endophagic filariasis, arbovirus and malaria vectors. *PLoS Neglected Tropical Diseases* **4(8)**: e773. doi:10.1371/journal.pntd.0000773

RBM (2011) *Business Investing in Malaria Control: Economic Returns and a Healthy Workforce for Africa*. Roll Back Malaria Progress and Impact Series 6. WHO, Geneva.

Roberts, D., Tren, R., Bate, R. and Zambone, J. (2010) *The Excellent Powder: DDT's Political and Scientific History*. Dog Ear Publishing, Indianapolis.

Schofield, C. J., Briceno-Leon, R., Kolstrup, N., Webb, D. J. T. and White, G. B. (1990). The role of house design in limiting vector-borne disease. In: Curtis, C. F. (ed.), *Appropriate Technology in Vector Control*. CRC Press, Boca Raton, Florida. pp. 187-212.

Schofield, C. J. and White, G. B. (1984) House design and domestic vectors of disease. *Transactions of the Royal Society of Tropical Medicine and Hygiene* **78**: 285-92.

Service, M. W. (1993) Community participation in vector-borne disease control. *Annals of Tropical Medicine and Parasitology* **87**: 223-34.

Smith, A., Obudho, W. O., Esozed, S. and Myamba, J. (1972) Verandah-trap hut assessments of mosquito coils with a high pyrethrin I/ pyrethrin II ratio against *Anopheles gambiae*, Giles. *Pyrethrum Post* **11**: 138-40.

Utzinger, J., Tozan, Y. and Singer, B. H. (2002) Efficacy and cost-effectiveness of environmental management for malaria control. *Tropical Medicine and International Health* **6**: 677-87.

Utzinger, J., Tozan, Y., Doumani, F. and Singer, B. H. (2002) The economic payoffs of integrated malaria control in the Zambian copperbelt between 1930 and 1950. *Tropical Medicine and International Health* **7**: 657-77.

WHO (1982) 'Manual on environmental management for mosquito control.' WHO, Geneva.

WHO (2004) 'Global strategic framework for Integrated Vector Management.' WHO/CDS/CPE/PVC/2004.10. WHO, Geneva.

WHO (2009) 'Guidelines for efficacy testing of mosquito repellents for human skin.' WHO, Geneva.

WHO (2011) 'Integrated Vector Management Handbook.' 80 pp. WHO, Geneva.

Xue, R.-D., Ali, A. and Barnard, D. R. (2003) Laboratory evaluation of toxicity of 16 insect repellents in aerosol sprays to adult mosquitoes. *Journal of the American Mosquito Control Association* **19**: 271-4.

Xue, R.-D., Barnard, D. R. and Ali, A. (2001) Laboratory and field evaluation of insect repellents as oviposition deterrents against the mosquito *Aedes albopictus*. *Medical and Veterinary Entomology* **15**: 126–31.

Yakob, L., Dunning, R. and Yan, G. (2010) Indoor residual spray and insecticide treated bed nets for malaria control: theoretical synergisms and antagonisms. *Journal of the Royal Society Interface* **8**. doi:10.1098/rsif.2010.0537

Yukich, J. O., Lengeler, C., Tediosi, F., et al. (2008) Costs and consequences of large-scale vector control for malaria. *Malaria Journal* **7**: 258.

Chapter 7

Other Insects – Flies, Cockroaches and Bed Bugs

Insects which spread disease mechanically by walking on surfaces with bacteria and then walk on humans are like malaria, most common in areas that suffer poverty. Insufficient money results in poor housing that is often close to where animals are kept, so insects transfer the disease organisms over quite a short distance. The key to avoid diarrhoea and similar diseases, caused by *Salmonella*, *Campylobacter* and other pathogens, is to maintain high standards of sanitation, so sources of food are not left on table tops and other surfaces where food is prepared. Chemical control should be needed only as a last resort. Where indoor residual spraying is used against Anopheline mosquitoes or triatome bugs, it will also provide some control of flies and cockroaches, but residual insecticide spray treatments should be avoided to reduce the risk of selecting resistant populations.

In good-quality houses, spot treatment with an insecticide may be required and this is often applied by using an aerosol can (otherwise known as a pressure pack), in which the insecticide dissolved in a solvent is kept under pressure. When the nozzle on the top of the can is depressed, a very fine spray is emitted and can be directed at surfaces under sinks or other areas where these insects are found. As an alternative to the aerosol can, it is still possible in some countries to purchase a 'Flit' gun, in which a small manually operated pump directs a jet of air over the end of a short tube that dips into a small container of insecticide. The jet of air sucks up liquid by a venturi action, so that the insecticide is blown a short distance as a spray of small droplets. A third technique is to distribute a low concentration dust formulation. The appropriateness of these techniques for various insect pests is discussed in the following sections.

Integrated Vector Management: Controlling Vectors of Malaria and Other Insect Vector Borne Diseases, First Edition. Graham Matthews.
© 2011 John Wiley & Sons, Ltd. Published 2011 by John Wiley & Sons, Ltd.

Flies

A major group of nuisance pests are the filth flies, as they are associated with surfaces contaminated with pathogens, such as faeces, food waste, animal manures and carrion. Most flies will be found when human dwellings are close to where animals are kept or there is a large area where refuse is dumped in a land fill site, but left uncovered. Thus keeping animal houses clean and free of faeces will significantly reduce fly populations. In many more affluent countries, waste food and other rubbish are kept in closed container bins, which are emptied regularly by the local authorities.

The most common fly worldwide is *Musca domestica* – the housefly. Other flies considered here are the face fly (*Musca autumnalis*), the stable fly (*Stomoxys calcitrans*), the false stable fly, (*Muscina stabulans*), the lesser housefly (*Fannia canicularis*), the green blow fly (*Lucilia sericata*), the blue blowfly (*Calliphora vicina*), the flesh fly (*Sarcophaga* spp.), and the face fly (*Musca sorbens*). Observations in The Gambia showed that more than 90% of the flies contacting eyes and spreading *Chlamydia trachomatis,* the cause of trachoma, were *M. sorbens.*

Where flies are a problem, it is important to maintain a surveillance programme by using sticky traps or other sampling techniques to establish the locality and severity of the fly problem. Various traps, some of which incorporate a pheromone attractant, are commercially available and can be used to control low populations of flies as well as indicate when other measures are needed.

In some areas, such as shops selling food, a trap with an ultra-violet light is used during daylight hours to attract any adult flies in the area. If used at night, they will collect many other insects as well as flies. The trap should be fitted with a sticky surface so that flies entering the trap are held by the glue and can be easily counted. Similar traps that electrocute flies landing on the trap are not recommended as the electrocuted fly may shatter, spreading small parts of the insect and any pathogens into the air.

In areas liable to attract flies, they can be prevented from entering houses by having a flexible screen over doorways or fitting a fan over the doorway that creates a curtain of air blowing any fly away from the door. Doors should also open outwards and automatically shut to reduce fly entry.

When a small number of flies do enter a dwelling, spot treatment with a hand operated pressure pack (Aerosol can) or flit gun directed at where the flies settle should give adequate control. In commercial establishments, apart from using an ultra-violet light trap, it is also possible to hang or attach to walls in selected places a granular bait, containing either spinosad or imidacloprid, affixed to a cardboard strip inside a mesh container. The mesh allows flies to enter, but prevents others contacting the insecticide. Other insecticides, mostly organophosphates, have been recommended for use in baits, but preference is given here to the recently introduced products. The bait contains sugar and a pheromone to attract the flies and a bitter agent to stop accidental ingestion by children or pets. A similar mixture of this bait with insecticide can also be used for spraying or painting on selected spots to attract flies that are exposed to a lethal deposit when landing on the spot. Although the deposit can remain effective for several weeks indoors, the residual deposit is confined to a limited surface area. Crack and crevice treatment is not recommended for fly control.

Refuse dumps

Apart from areas with domestic animals, the main source of breeding of flies is where waste food and other rubbish are collected. Larval control is not always feasible, where additional waste may be added daily and so buries any larvicide treatment or dilutes its effect. The main larvicides that should be used are insect growth regulators and other types of insecticide; especially pyrethroids must be avoided and used only against adult flies. Apart from the need to reduce selection for resistance, the application of Insect Growth Regulators (IGR) should have less impact on natural predators of fly larvae, compared with using non-selective insecticides.

Application of larvicides is usually at a high volume ($250\,ml/m^2$) to get the spray to saturate the area where breeding is taking place. Various types of manually carried compression, lever operated or motorised sprayers can be used for larvicide application, although motorised hydraulic sprayers can deliver a spray at a higher pressure. Several targeted applications may be required during periods with large numbers of flies and during epidemics of fly-borne diseases, such as dysentery.

Larvicides recommended by WHO for housefly control are shown in Table 7.1.

Space treatments

If infestations are severe, a space treatment using a thermal or cold fog may be needed. As flies emerge from breeding sites daily, sequential treatments may be needed over a 1-2 week period in some situations, but normally 1-2 applications should be sufficient if timed in relation to routine surveillance sticky trap data. Overuse of chemicals applied as space sprays should be avoided to reduce selection of a resistant fly population. Portable equipment may be used around waste dumps where vehicle access is limited. Treatments need to be applied when there is little or no wind, to avoid the fog dispersing

Table 7.1 Larvicides for fly control.

Insecticide	Dosage (g ai/m²)
Diflubenzuron	0.5–1.0
Cyromazine	0.5–1.0
Pyriproxifen	0.05–0.1
Triflumuron	0.25–0.5

Table 7.2 Some recommended insecticides for space treatments. Dosage is for cold fogs. For some insecticides, the dosages applied in thermal fogs may be slightly different.

Insecticide	Type	Dosage (g ai/ha)
Chlorpyrifos methyl	OP	100–150
Malathion	OP	672
Pirimiphos methyl	OP	250
Bioresmethrin	Pyrethroid	5–10
α-Cypermethrin	Pyrethroid	2–5
Cyphenothrin	Pyrethroid	5–10
Deltamethrin	Pyrethroid	0.5–1.0
Esfenvalerate	Pyrethroid	2–4
Etofenprox	Pyrethroid	10–20
λ-Cyhalothrin	Pyrethroid	0.5–1.0
Permethrin	Pyrethroid	5–10
λ-Cyhalothrin + tetramethrin + piperonyl butoxide	Synergised pyrethroid	0.5 + 1.0 +1.5
Deltamethrin – Bioallethrin + piperonyl butoxide	Synergised pyrethroid	0.3–0.7 + 0.5–1.3 + 1.5

too quickly. Normally about 0.5–20 l/ha of spray is applied with a cold fogger and up to 50 l/ha with a thermal fogger. WHO has recommended a number of pyrethroid insecticides, used alone or as mixtures, some of which are synergised with piperonyl butoxide for application in fogs to control flies. Selected low toxicity organophosphate insecticides, such as chlorpyrifos methyl, malathion and pirimiphos methyl, may also be used to break the cycle of using pyrethroids. Table 7.2 shows selected examples.

Mist treatments

Instead of cold fogs, mist applications may be applied. Populations of *M. sorbens* were reduced by using a knapsack mist blower applied around buildings and this reduced the number of trachoma infections of the eye and also reduced diarrhoea transmitted by *M. domestica* (Emerson et al., 2000). In chicken houses, portable sprayers (Figure 7.1) and small hand carried equipment with a rotary nozzle have also been used to control flies.

Cockroaches

There are a few species of cockroach that invade dwellings and other buildings, such as restaurants and hospitals. They are found especially in warm areas and in particular kitchens, so can spread enteric diseases, such as *Salmonella*, *Shigella* and *Escherichia coli*, to food. These pathogens have been identified from samples taken from the gut and faeces of cockroaches.

Figure 7.1 Treating a poultry house for fly control (photo: Clive Boase).

Some people become sensitised to cockroaches when living in dwellings with a large population of them. This sensitivity to cockroach allergens has been associated with asthma. Young children in families with a history of atopy are reported to suffer from recurrent wheezing (Litonjua et al., 2001).

The most widely distributed are the German cockroach (*Blatella germanica*), the American cockroach (*Periplaneta americana*) and the Oriental cockroach (*Blatta orientalis*). Other species may be locally important. Generally the public considers that they are a health hazard that has led to most (90%) of the pesticides applied by those living in apartments in New York being aimed at cockroaches (Whyatt et al., 2002). Phasing out the use of the more toxic organophosphate (OP) insecticides, chlorpyrifos and diazinon, together with improved cleaning as part of IPM, fewer insecticide treatments with less toxic chemicals was possible (Williams et al., 2008). In a comparison between IPM (clearing apartments of cockroaches using vacuum cleaners and sticky traps and using baits) with use of baits alongside the normal cleaning in apartments, showed that the IPM method resulted in fewer cockroaches and was more sustainable, but costs more to implement (Wang and Bennett, 2006).

A key part of cockroach control is the removal of food, garbage, harbourage sites and debris, so that the cockroaches have no food or places to hide. Check to see that there is no water leaking from pipes, so as to prevent cockroaches having access to water. General cleanliness in areas likely to attract cockroaches is a first step. Where there are cracks in walls or other harbourages, structural repairs are needed, as this will reduce hiding places and expose any cockroaches to any insecticide treatment that has to be carried out.

Traps

Cockroaches are active at night, so their presence can be detected by using traps with a food and pheromone attractant. These can be placed in areas where cockroaches might be expected. There are different trap designs that can vary in their effectiveness (Stejskal, 1998), although sticky traps are generally used. Use of traps is crucial in surveillance to:

- determine which species is present;
- detect where cockroaches are active, so that treatments can be localised;
- decide on the control needed;
- adjust the level of control needed in relation to the trap catches; and
- check on effectiveness of control measures. If there is more than one cockroach per trap per night, any existing control measure may need to repeated or changed by using a different bait, treating different places or increasing the number of places treated with bait. Alongside any chemical control, further emphasis is needed on thorough cleaning, so that waste food is not attracting the cockroaches.

On aircraft, a pheromone trap can be placed behind a panel in the galley so that on routine maintenance schedules, the effectiveness of previous treatments can be checked. However, traps are not effective at controlling the pest. Where cockroaches are found in blocks of apartments, it is important to get the tenants, landlords and all other stakeholders in the area to co-operate, so action can be taken throughout the building. In some cases, an aerosol spray may be used to flush cockroaches from behind panels.

Sprays

Many people use aerosol (pressure pack) sprays to control cockroaches in their houses. However, with resistance to many of the older and frequently used insecticides now detected in many situations, a comparison was made between sprays of propoxur and deltamethrin + allethrin and gels containing imidacloprid or fipronil. A single application of fipronil gel reduced the infestation by 96.8% within 12 weeks, while the imidacloprid gel achieved a 90.9% reduction based on visual counts (Figure 7.2). The pyrethroid spray failed to achieve adequate control and the propoxur spray, although it gave better control in the first week, was not sustained after 8 weeks (Agrawal et al., 2010).

Sprays using a compression sprayer and cone or solid stream (pin-jet) nozzle have been applied for treating surface and 'Crack and Crevice' treatments in areas where cockroaches have been detected. Such treatments may be needed if very large infestations are found. On aircraft, specialist pest control operators are trained so that panels in the galley and toilets of the aircraft can be moved to allow sprays to be applied behind the panel. Routine treatments are carried out when the aircraft is undergoing routine maintenance. A number of insecticides are used in rotation in an effort to avoid selecting for resistance.

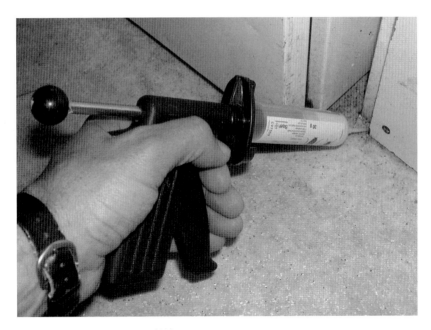

Figure 7.2 Applying insecticide gel with special applicator for cockroach control (photo: Clive Boase).

Baits

The use of baits to control cockroaches is now recommended instead of sprays, as this minimises where treatments are needed and avoids any surfaces where food may be prepared. Baits can be placed in areas under sinks, behind or along pipes and on top of cupboards inaccessible to children.

Some baits are formulated as a gel, in which an insecticide, such as imidacloprid or fipronil, is mixed with one or more attractants. The speed of action will depend on the insecticide used. Some can achieve a rapid kill (e.g. imidacloprid), quick (e.g. fipronil) or slow action (e.g. hydramethylnon) (Stejskal et al., 2004). Slow action allows cockroaches that have eaten bait to go back to hiding places where other cockroaches may eat those that die. Such baits are more effective against the early instar nymphs. Hydramethylnon in a bait evaluated in infested cafeterias in Thailand reduced the number caught in traps by more than 90% in a week and remained effective for up to 3 months post-treatment (Sitthicharoenchai et al., 2006). However, within Europe, the use of hydramethylnon is no longer supported under the Biocide Product Directive. Other experimental studies in cockroach-infested kitchens have shown that a 0.25% indoxacarb gel bait significantly reduced German cockroaches by 74% after 3 days and by 95% in 2 weeks (Appel, 2003). Baits should not be repellent, as this will tend to cause them to scatter to different areas. In sensitive areas, the baits can be enclosed in a mesh cage. Baits can also contain a bitter agent to discourage other animals from eating the bait.

Attracting the cockroaches to a palatable bait allows more selective placement. A bait packaged in a tube can be extruded and applied as a thin line or placed as discrete spots (0.6 cm diameter), so there is about 1 spot every $0.1 m^2$. A 0.6-cm diameter bead will deliver approximately 0.1 g of the bait. Thus, a 30 g tube of gel applied as 0.1 g beads should treat 300 m². The amount applied per unit area should be according to the product label and adjusted if necessary according to the extent of the infestation. Special applicators are available to facilitate applying the gel. Such applicators simply displace gel by applying pressure on a piston within the gel tube, so that the gel is extruded through a nozzle.

Where there are any cracks and crevices in walls, a small quantity of the gel (about the size of a pea) can be injected into the cracks. The latter is important where a crack leads to a cavity or void in the wall, where the cockroaches may be breeding. Similarly bait may need to be placed behind any boxes or cupboards adjacent to a wall. The amount of bait recommended is usually an amount per unit area, but this may need to be adjusted where a large number of cockroaches are present. Sprays should not be applied in areas where gel baits are applied, as the toxic surface may affect the behaviour of the cockroaches and limit the amount of bait taken (Stejskal and Aulicky, 2006).

A traditional chemical for cockroach control is boric acid. It is slow acting, so may also be mixed with natural pyrethrins for a quicker kill. It is also used in baits. A range of different insecticides is needed, as cockroaches have been shown to become resistant to some of the older insecticides, such as propuxor and chlorpyrifos, used in baits (Lee, 1998) as well as sprays.

Baits should not be placed on hot surfaces or in exposed places, which may be washed frequently, as a single application of bait can last for several days. Baits can be used outdoors with placement around any pipes or other openings on the external walls, through which cockroaches can enter a house.

Bed bugs

Blood-feeding true bugs of the Cimicidae, notably *Cimex lectularius*, the common bed bug, have been persistent pests on man. In the tropics, *C. hemipterus* and *C. rotundus* occur. Reports of bed bugs decreased following widespread use of organochlorine insecticides in homes, but since about 2000, there has been a significant increase in the number of reports of bed bug infestations. Infestations have been reported in many different situations, ranging from low-cost hostels to luxury hotels (Harlan, 2007). Each nymphal stage needs a blood meal to develop into the next stage and adults may feed every 3–5 days over a lifespan of 6–12 months, although they can survive very long periods without feeding. Bites can cause itching due to the saliva, which contains enzymatically active proteins that cause an allergic reaction. If an infestation is left uncontrolled, it has been reported to cause anaemia (Pritchard and Hwang, 2009). Dust from bed bug infestations that is airborne can cause allergic reactions and may result in bronchial asthma. Transmission of more than 40 human diseases has been attributed to bed bugs, but there is little evidence that they are vectors of communicable disease (Goddard and de Shazo, 2009).

Control of bed bugs starts with an inspection to consider the extent of the infestation, which will enable appropriate decisions regarding the control

Figure 7.3   Treating a mattress with bendiocarb for bed bug control (photo: Clive Boase).

measures needed. With public concern about the use of insecticides, emphasis is on integrating cultural methods. If chemical control is used (Figure 7.3), further use of cultural methods is generally needed so that the problem does not recur.

Often bed bugs will have colonised mattresses and other bed linen, cracks and crevices in the bed and adjacent walls, usually within 2 m of a bed, so initially the area needs to be cleaned by vacuuming the area, while bed linen is washed at temperatures above 50°C for at least 20–30 minutes. This contrasted with data that showed no mortality occurred when adult bed bugs were exposed to 39°C for 4 hours (Pereira et al., 2009). Cracks and crevices can be sealed and/or treated with steam. The steam needs to be emitted through a tube, the tip of which is 2.5–4 cm from the target surface, to ensure a sufficiently high temperature without wetting the surface. It is directed especially at the seams and moved along at about 30 cm in 15 seconds. An alternative to heat treatment is to freeze infested items for a minimum of 4 days. Mattresses can be enclosed in a plastic cover which is sealed, preferably with a high concentration of CO_2. Dry ice can be used to obtain CO_2 within the sealed space.

Table 7.3 Suggested insecticides for bed bug control.

Insecticide	Type	Concentration (g/l or g/kg)
Bendiocarb	Carbamate	2.4
Flufenoxuron	IGR	0.3
Methoprene	IGR	0.9
Pirimiphos methyl	OP	10
Chlorfenapyr	Pyrrole	0.5
α-Cypermethrin	Pyrethroid	0.3–0.6
Cyphenothrin	Pyrethroid	0.5–1.0
λ-Cyhalothrin	Pyrethroid	0.03
Permethrin	Pyrethroid	1.25

If chemical control is needed, it is usually by applying a residual insecticide to all the areas that have been infested. A compression sprayer fitted with a constant flow valve and using a fan nozzle can be used to treat floors and walls (see Chapter 2 for further details of indoor residual spraying), but a pin-jet is needed to penetrate spray into crack and crevices. Use of cone nozzles is not recommended. The volume of spray can be between 20 and 40 ml/m², with the higher volume used on absorbent surfaces. The effect of any treatment should be monitored, as nymphs may emerge within 4–6 weeks of spraying if eggs were unaffected by the treatments. Residual spraying should be done early in the day, so that deposits have dried thoroughly before the rooms are occupied at night.

Many insecticides have been used, including organophosphates and pyrethroids, but they are not always effective due to resistance, with resistance to pyrethroids already widespread in the USA (Romero et al., 2007). Natural pyrethrins, synergised with piperonyl butoxide, are effective and can be useful in revealing where the beds bugs are by flushing out them out from their hiding places. Behavioural factors were shown in a study where bed bugs were able to avoid surfaces treated with deltamethrin but not deposits of chlorfenapyr (Romero et al., 2009). Insect growth regulators have also been shown to be effective and are sometimes used together with a pyrethroid, for example, flufenoxuron + α-cypermethrin. Silica gel as a dust is also recommended. Dusts are normally only used in voids within walls, as they can be readily vacuumed by the occupant of the room. Recommended insecticides are shown in Table 7.3, with different chemicals used if an area requires further treatment to avoid excessive use of insecticides with the same mode of action. The micro-encapsulated formulations (CS) are generally more effective.

A high-rise apartment block, with 16 bed bug infested apartments, was the site of a trial that investigated the use of diatomaceous earth as a dust in comparison with 0.5% chlorfenapyr spray. Initially steam was applied to bed frames, floors under the beds, perimeter of the floor, sofas, and other furniture in the infested areas. Mattresses were encased and bed linen laundered. Bed bug interceptors, consisting of two plastic bowls, with the smaller one placed inside the larger one, formed a trench which contained 40 ml of 50% ethylene glycol to kill any bed bugs that climbed a piece of fabric glued to the outside of the larger bowl. A small amount of diatomaceous earth and talc was put in the smaller bowl, before the legs of the bed and chairs were placed in the small bowl of the 'bed bug interceptors'. These traps were monitored twice a week for 10 weeks, and apartments re-treated when necessary. After 10 weeks, both treatments had eradicated bed bugs from 50% of the apartments, but bed bugs continued to re-infest apartments from hallways, partly due to lack of concern by the residents (Wang et al., 2009). The trial illustrates the problem of control if the whole area is not treated, as bed bugs can readily re-infest an area that is not protected by some form of residual treatment. Clearly educating those who suffer an infestation is needed, so that they understand how to try and avoid re-infestations occurring.

In Australia, there is a manual on bed bug control issued by the Australian Environmental Pest Managers Association (Doggett, 2010).

Insecticide treated bed nets will also provide some protection from bed bug bites.

Conclusion

Although the insects considered in this chapter are not directly concerned with disease transmission, their behaviour can result in spread of disease or discomfort to individuals. Control may be partially achieved by controls aimed at vectors, but special attention to their control, including good hygiene, will reduce their impact on man.

References

Agrawal, V. K., Agarwal, A., Choudhary, V., et al. (2010) Efficacy of imidacloprid and fipronil gels over synthetic pyrethroid and propoxur aerosols in control of German cockroaches (Dictyoptera: Blatellidae). *Journal of Vector Borne Diseases* **47**: 39-44.

Appel, A. C. (2003) Laboratory and field performance of an Indoxacarb bait against German cockroaches (Dictyoptera: Blatellidae). *Journal of Economic Entomology* **96**: 863-70.

Doggett, S. L. (2010) *A Code of Practice for the Control of Bed Bug Infestations in Australia*, 3rd edn. Australian Environmental Pest Managers Association Ltd.

Emerson, P. M., Bailey, R. L., Mahdi, O. S., Walraven, G. E. and Lindsay, S. W. (2000) Transmission ecology of the fly *Musca sorbens*, a putative vector of trachoma. *Transactions of the Royal Society of Tropical Medicine and Hygiene* **94**: 28-32.

Goddard, J. and de Shazo, R. (2009) Bed bugs (*Cimex lectularius*) and clinical consequences of their bites. *Journal of the American Medical Association* **301**: 1358-66.

202 ■ **Integrated Vector Management**

Harlan, H. J. (2007) Bed bug control: challenging and still evolving. *Outlooks on Pest Management* **18**: 57–61.

Lee, C. Y. (1998) Control of insecticide-resistant German cockroaches, *Blattella germanica* (L.) (Dictyoptera: Blattellidae) in food-outlets with hydramethyl non-based bait stations. *Tropical Biomedicine* **15**: 45–51.

Litonjua, A. A., Carey, V. J., Burge, H. A., Weiss, S. T. and Gold, D. R. (2001) Exposure to cockroach allergen in the home is associated with incident doctor-diagnosed asthma and recurrent wheezing. *Journal of Allergy Clinical Immunolgy* **107**: 41–7.

Pereira, R. M., Koehler, P. G., Pfiester, M. and Walker, W. (2009) Lethal effects of heat and use of localized heat treatment for control of bed bug infestations *Journal of Economic Entomology* **102**: 1182–8.

Pritchard, M. J. and Hwang, S. W. (2009) Severe anaemia from bed bugs. *Canadian Medical Assocation Journal* **181**: 287.

Romero, A., Potter, M. F., Potter, D. A. and Haynes, K. F. (2007) Insecticide resistance in the bed bug: a factor in the pest's sudden resurgence? *Journal of Medical Entomology* **44**: 175–8.

Romero, A., Potter, M. F. and Haynes, K. F. (2009) Behavioral responses of the bed bug to insecticide residues. *Journal of Medical Entomology* **46**: 51–7.

Sitthicharoenchai, D., Chaisuekul, C. and Lee, C.-Y. (2006) Field evaluation of a hydramethylnon gel bait against German cockroaches (Dictyoptera: Blattellidae) in Bangkok, Thailand. *Medical Entomology and Zoology* **57**: 361–4.

Stejskal, V. (1998) Field tests on trapping efficiency of sticky traps for *Blatta orientalis* and *Blatella germanica*. *Anzeiger fur Schadlingskunde* **71**: 17–21.

Stejskal, V., Lucas, J. and Aulicky, R. (2004) Speed of action of 10 commercial insecticidal gel-baits against the German cockroach, *Blattella germanica*. *International Pest Control* **46**: 185–9.

Stejskal, V. and Aulicky, R. (2006) Can the size of a bait drop affect the efficacy of German cockroach control? *International Pest Control* **48**: 196–8.

Wang, C. and Bennett, G. W. (2006) Comparative study of integrated pest management and baiting for German cockroach management in public housing. *Journal of Economic Entomology* **99**: 879–85.

Wang, C., Gibb, T. and Bennett, G. W. (2009) Evaluation of two least toxic integrated pest management programs for managing bed bugs (Heteroptera: Cimieidae) with discussion of a bed bug intercepting device. *Journal of Medical Entomology* **46**: 566–71.

Whyatt, R. M., Camann, D. E., Kinney, P. L., et al. (2002) Residential pesticide use during pregnancy among a cohort of urban minority women. *Environmental Health Perspectives* **110**: 507–14.

Williams, M. K., Barr, D. B., Camann, D. E., et al. (2008) Residential pesticide use patterns among an inner-city cohort in New York City and the impact of an intervention to reduce pesticide use and exposure. *Epidemiology* **19**: Suppl S41.

Chapter 8

Looking Ahead

New insecticides?

Instead of controlling the vectors of diseases, such as malaria, many feel the aim should be the eradication of the disease. However, this was the objective in the early days of DDT, but was abandoned not only because of the costs of maintaining control over a prolonged period, but because the mosquito vectors became resistant to DDT. The same dilemma is now confronting vector control with so much reliance on use of pyrethroid insecticides, with sufficient persistence to have a long residual effect, as the modern replacement for DDT. These insecticides have been used extensively to control agricultural pests, so many of the problems of resistance occur where the vectors are exposed not only to the spray deposits in houses but also in the general environment.

Ideally there would be a certain insecticide kept solely for vector control, but the huge costs of developing a new insecticide, invested over a period of 7-10 years to obtain all the data needed for registration, have to be recovered from using the chemical as extensively as possible in agriculture. Thus, public health insecticides are almost always a spin-off from developments for agriculture. This has meant that few effective insecticides are available for programmes to control malaria and dengue (Zaim and Guillet, 2002). Most of the major R&D agrochemical companies have not screened for activity against public health pests, due the relatively small market and insufficient financial return. To overcome this problem, the Innovative Vector Control Consortium was established with the Bill and Melinda Gates Foundation funding, to develop a portfolio of technological tools to bring vector control into the twenty-first century (Hemingway et al., 2006). Some public health pesticides are being lost as the manufacturers have been unwilling to re-register products due to the high cost of collecting additional data, for example to test for endocrine disruption. In the USA, this loss of many insecticides previously registered for vector control has led the American Forces Pest Management Board (AFPMB)

Integrated Vector Management: Controlling Vectors of Malaria and Other Insect Vector Borne Diseases, First Edition. Graham Matthews.
© 2011 John Wiley & Sons, Ltd. Published 2011 by John Wiley & Sons, Ltd.

to link with the United States Department of Agriculture Research Service and the IR-4 Project, set up to help obtain registration of pesticides for minor agricultural crops, in an effort to facilitate development and registration of public health pesticides (Malamud-Roam et al., 2010).

The agrochemical industry has changed in terms of its criteria for new insecticides and has not been so concerned about persistence of surface deposits, but more about the ability of the plant to take up the chemical and distribute it through the plant. Thus, a number of new insecticides have been developed for controlling insects, such as aphids and other pests that feed by sucking plant tissues. The agrochemical industry has also invested in genetic engineering and has incorporated genes that express the toxin from *Bacillus thuringiensis* to control lepidopteran larvae (and in some cases other insect groups), while relying on systemic insecticides to control sucking pests. Vector control does need persistent insecticides that remain effective over a long period to avoid having to retreat a house or other building frequently. This is in contrast with agriculture, where a pest may invade a crop for a specific period and plant growth dilutes the impact of any individual spray, so although a photo-stable insecticide is needed, the period of persistence need only extend for 7-10 days in most cases.

Can insecticides with new modes of action be developed?

Paul et al. (2006) reported on some tests against *Aedes aegypti* adults and larvae with new insecticides developed for use in agriculture. The four insecticides, highly or moderately toxic to larvae, were chlorfenapyr, hydramethylnon, imidacloprid and indoxacarb. Pyriproxyfen was also shown to be effective against larvae. Activity against adults was shown with diafenthiuron and chlorfenapyr as well as imidacloprid, which was strongly synergised by piperonyl butoxide (PBO). So far, use of some of these as larvicides has been reported, but not against adults. Chlorfenapyr has been compared as an indoor residual spraying (IRS) treatment in experimental huts in Tanzania with a pyrethroid alpha cypermethrin and shown to have potential, especially where pyrethroid resistant mosquitoes occur (Oxborough et al., 2010). Similar studies have been reported in Benin (N'Guessan et al., 2009). N'Guessan et al (2007a) also have reported on data using indoxacarb with no excito-repellence in adults, but is slower acting.

Some insecticides, which have been used in agriculture over a long period, are now available as new micro-encapsulated (CS) formulations and are being tested as adulticides. Thus, the organophosphate insecticide pirimiphos methyl and carbamate bendiocarb are being tested for IRS. Chlorpyrifos methyl has been used in experimental bed nets.

The pyrethroids were developed as a result of a search for a more photo-stable version of the botanically derived natural pyrethrins. The latest new insecticide introduced into agriculture is an anthranilic diamide that acts on the ryanodine receptors. This, it is a similar development of copying a botanical insecticide, in this case ryania, which is extracted from a plant in South

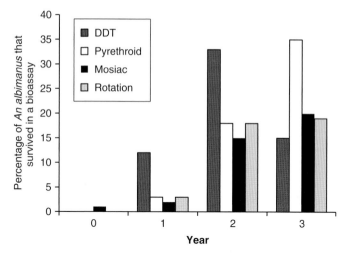

Figure 8.1 Increase in proportion of resistance to pyrethroids over three years with different treatment regimes. Villages were assigned four treatments: 1) two sprays per annum of DDT; 2) three applications per annum of a pyrethroid; 3) a spatial mosaic of organophosphate and pyrethroid; or 4) an annual rotation of organophosphate, a pyrethroid and a carbamate (data redrawn from Penilla et al., 2007).

America. Whether this or other chemicals acting on different receptors will be developed for vector control is at present unknown.

As DDT was also repelling mosquitoes from houses, perhaps looking for this characteristic in new compounds should not be ignored. It is hoped that advances in genomics, specifically the sequencing of the *Anopheles gambiae* genome, will enable scientists to identify new target sites that could lead to the development of new insecticides, which might include specific activity against dipteran vectors and be sufficiently persistent on walls to minimise the number of applications needed.

However, Read et al. (2009) have argued that any new insecticide developed for controlling *Anopheles* mosquitoes should be targeted at old mosquitoes, and ideally those infected with the malarial parasite, as this could provide effective malaria control while only weakly selecting for resistance.

Insecticide resistance

Resistance is defined by the Insecticide Resistance Action Committee (IRAC) as the selection of a heritable characteristic in an insect population, which results in the repeated failure of an insecticide product to provide the intended level of control when used as recommended. Thus, differences in susceptibility apparent in laboratory bioassays may not necessarily constitute resistance, if the difference does not result in a change in the field performance of the insecticide (http://www.irac-online.org/) (Figure 8.1).

A crucial problem with vector control is the current reliance on pyrethroids (Ranson et al., 2011). Selection for resistance can occur very rapidly with extensive use of insecticides with the same mode of action (Figure 8.1). Long-lasting insecticidal bed nets (LLIN) and durable wall linings present deposits that remain active over long periods, so that selection for resistance is continuous. Skovmand (2010) has argued that IRS and textiles hung on walls should not be treated with the same insecticide, as this will render the nets less effective due to resistance.

High mortality of free-flying *An. gambiae* has been reported, where people were sleeping under pyrethroid treated nets in an area where there is a high frequency of the *kdr* resistance gene (Darriet et al., 2000; Asidi et al., 2004). As pointed out by Curtis (2010), the reduced irritability of mosquitoes with the *kdr* gene suggests that these mosquitoes remain in contact with the pyrethroid net longer and eventually pick up a lethal dose. Also Rajatileka et al. (2011) reported higher mortality occurred with 14-day old mosquitoes than the 1-3 day old unfed mosquitoes used in most bioassays. However, more recent data (N'Guessan et al., 2007b) indicated a failure to control in Benin. Curtis (2010) advocated using bendiocarb or pirimiphos methyl as adulticides and pyriproxifen as a larvicide to eradicate if possible the resistant *An. gambiae*.

Until new insecticides based on novel chemistry can be developed, there needs to be an improved policy regarding the use of existing insecticides for vector control. Prior to the development of genetically modified (GM) crops, the same problem of widespread resistance due to overuse of pyrethroids had to be tackled. This was most clearly developed in Australia, where the pyrethroids were used extensively to control bollworm larvae on cotton, so a programme was introduced that limited their use on any crop in a single year to a period of about 1 month. This was designed to allow spraying of cotton with a pyrethroid during the main period of *Helicoverpa armigera* infestation, with the assumption that by using other types of insecticide in the other 11 months, any selection for resistance to pyrethroids would have been lost by the following year. Alongside this approach, a synergist PBO was added to pyrethroid insecticide sprays to enhance insecticide efficacy in insects where esterase resistance mechanisms occur.

Another pragmatic resistance management programme, referred to earlier in Chapter 2, was developed in Zimbabwe, where spider mites were resistant to the organophosphate dimethoate following its use on cotton and also vegetables throughout the year. In this programme, the country was divided into three sections in which acaricides with different modes of action were rotated, so that one mode of action was used for only 2 years in one section of the country. Unused stocks of chemical were then transferred to the next section in which it could be used.

Using DDT for IRS did lead to mosquitoes with resistance to it and history is repeating itself with the use of pyrethroids, as there is already evidence of resistance in some areas where pyrethroids are used in both vector control and agriculture. It is possible that the selection of resistance may have been initiated by exposure of the insecticides to larvae, especially where water became contaminated with insecticide. Thus, ideally insecticides needed for vector control should not be used in agricultural spraying and the insecticides that are used for vector adulticiding should not be used also as larvicides. If organophosphate insecticides in new micro-encapsulated formulations provide greater persistence on wall surfaces, it seems that there should be a management policy to rotate their use with pyrethroid and DDT, where its use in vector control can be managed, so that selection for resistance to one mode of action is less intense. As resistance can also occur with a rotation of insecticides used in IRS (Figure 8.1), it is important that ideally there should be at least three different modes of action, with each used for 2 years within its allocated area before being changed and that the programme must include larviciding, using completely different insecticides and/or drainage to reduce larval survival. However, ultimately choice of insecticide will depend on the outcome of susceptibility tests.

Bio-pesticides

A number of fungi have been considered as possible myco-insecticides for mosquito control. Most studies have been directed at their use as larvicides, for example *Lagenidium giganteum* has caused high mortalities in mosquito populations in many laboratory studies, especially in *Culex and Mansonia* spp., but the fungus is not effective for mosquitoes in brackish or organically rich aquatic habitats. Following the recognition of *B. thuringiensis israelensis*, further work on myco-insecticides was limited, but a review by Scholte et al. (2004) has considered that Hyphomycetes, such as *Metarhizium* and *Beauvaria*, can be used in vector control, as these fungi can be mass produced. Thomas and Read (2007) pointed out that small differences in mosquito longevity after an infective blood meal can still affect the transmission of the malaria parasite, so the slower action of a biopesticides could be as effective as a rapid kill. Blanford et al. (2005) had shown that with a rodent malaria model, the number of blood-fed mosquitoes able to transmit malaria was reduced by a factor of about 80, following exposure to surfaces treated with a fungal entomopathogen. Mortality increased, especially around the time of sporozoite maturation, and infected mosquitoes were less likely to feed.

Farenhorst et al. (2009) have shown a significant increase in mortality of mosquitoes pre-infected with *B. bassiana* or *Metarhizium anisopliae* after exposure to an insecticide, when compared with uninfected control mosquitoes. This suggests that the use of fungal biopesticides can complement existing vector control measures by providing products for use in resistance management strategies. Thus, an integrated fungus-insecticide control can potentially reduce malaria transmission in areas where pyrethroid insecticides are no longer sufficiently effective due to resistance (Farenhorst et al., 2010).

A major problem of transferring laboratory data to field use was illustrated by the development of *Metarhizium acridum* for locust control, in which careful development work of a suitable formulation to be applied at ultra-low volume to suit the field requirements of locust control was a key factor in implementation its use.

Studies on the effect of *Metarhizium* on adult female *An. gambiae* have shown that they produce fewer eggs and a shorter period of fecundity, indicating that the fungus could be an important biopesticide for vector control (Scholte et al., 2006). With vector control insecticides being applied in dwellings, one concern was the risk of entomopathogens to human health. Darbro and Thomas (2009) have reported that although airborne conidia were detected immediately after treatment, with concentrations of $7,000/m^3$, the number decreased during the next 48 hours to 500 conidia/m^3, and *Metarhizium* conidia decreased from only 2% of total visible particulate matter to 0.1% within 2 days. With mud houses and thatched roofs, the brief increase in aeropathogens due to a *Metarhizium* application to surfaces in a house is unlikely to cause a health risk.

Success of using a biopesticide will be dependent on how the conidia are applied, so that they are readily picked up by mosquitoes. Quantitative real time polymerase chain reaction (PCR) techniques have been developed, which could assist optimisation of myco-insecticides, and in initial tests confirmed that exposure to higher densities of conidia resulted in significantly greater pick-up by mosquitoes, leading to more rapid mortality (Bell et al., 2009) In addition to the possibility of using them by spraying wall surfaces, they could be applied to nets but an attract and infect technique could be by using clay pots used for storing water, which are an attractive resting site for mosquitoes. In an experiment, the inside of pots were sprayed initially with 30 ml of oil before being left to dry for 2 hours and were then sprayed with a suspension of conidia. Over 90% of *An. gambiae* were infected in cage trials where the pots were treated with $2\text{-}4 \times 10^{10}$ conidia/m^2 and significantly reduced their longevity (Farenhorst et al., 2008), indicating that the point source dispersal system warranted examination as part of an integrated vector management programme. Thus, any mosquitoes that survived entering a house treated by IRS or with LLINs could be affected by choosing a treated water pot as a resting site. Limited use of *Metarhizium* in this way may overcome the concerns about the higher cost of a myco-insecticide compared with chemicals, but also could reduce the survival of mosquitoes exposed to pyrethroid sprayed deposits or treated bed nets.

Another microbial technique showing promise is the release of *Ae. aegypti* mosquitoes infected with the endosymbiont *Wolbachia pipientis*, which reduces their ability to carry the dengue virus (Turley et al., 2009). Releases are planned in Australia and Vietnam.

Spray technology

The technology of using manually operated portable compression sprayers with a specific fan nozzle is now regarded by many as 1950s technology, yet little or no resources have been provided to determine whether an alternative method could be used in IRS.

Electrostatic spraying?

Studies were carried out by Chadd (1990), who investigated the use of an electrostatic system in which the electrostatic charge was used to create the spray droplets of an oil based formulation. The 'Electrodyn' system had been developed to treat agricultural crops at ultra-low volume, but by a modification using deflectrodes, the charged droplets could be directed at wall surfaces. One perception against the technique was the unacceptability of the odour of a solvent used to dissolve the pyrethroid insecticide, as it had to be mixed with an oil to provide the appropriate resistivity of the spray solution. The technique was not supported due to the limited number of actives that could be formulated for use at ultra-low volume with the 'Electrodyn' sprayer in agriculture.

Alternative electrostatic spraying systems, which use an air flow to direct droplets away from the nozzle, are being developed for IRS. A potential problem with air assisted spraying indoors is that the air flow is likely to result in more airborne droplets, which will sediment on the floor. However, spray on the floor must be avoided, especially if children are crawling on the floor. Another problem is that if an air-assisted system is used, the power requirements are considerably greater to move air than to spray liquid.

Different sprayers?

One of the problems with IRS is that it is regarded as essential that the wall surfaces are completely covered with insecticide, otherwise any mosquito repelled by a deposit would rest on a part of the wall that has not been sprayed. The assumption is that trained operators will be able to direct a fan spray in a regular pattern without any significant gaps. The compression sprayer was adopted because once the tank was pressurised, the operator only has to direct the nozzle. However, the tank has to be re-pressurised each time it is refilled and this requires considerable effort. Since the 1950s, other types of sprayer have been developed, but not assessed for IRS. In particular, as mentioned in Chapter 2, it is now possible to purchase knapsack sprayers with a small electric pump operated with a re-chargeable battery. The battery could be charged overnight and as the drudgery of pumping is eliminated, the operator can, as with the compression sprayer, devote attention to the direction of the nozzle.

Different nozzles?

Another problem is that not all the wall surfaces are highly absorbent, so applying the original recommendation of $40\,ml/m^2$ results in liquid running down certain types of wall. Some have recommended changing from the 8002 nozzle to an 8001 one to halve the volume, but as the standard flat fan nozzle has an elliptical orifice, the smaller orifice is more likely to become blocked, especially if water quality is poor and the spray concentration is doubled to achieve the same dosage of insecticide. Many new nozzles have been designed over the last 40 years and some may be as good as the hardened stainless steel standard fan nozzle 8002 or the even spray 8002E nozzle currently recommended. Modern manufacturing techniques can now produce ceramic nozzle tips that are less expensive than hardened stainless steel, but the key question is whether an acceptable flat fan distribution can be achieved with a round orifice that will be less prone to blockages at lower flow rates. Deflector nozzles are available that should be investigated together with the possibility of using wider spray angles, for example, 110 degrees compared to 80 degrees, to achieve a wider swath on walls.

Using a paint

Paints incorporating an insecticide were originally developed for the control of pests of stored produce that rested on warehouse walls, but recently the idea of painting non-porous walls of houses has been investigated for mosquito and Triatoma control (Amelotti et al., 2009). Mosqueira et al. (2010a, b) have reported on the evaluation of an insecticide paint, Inesfly 5A IGR™, which contains two organophosphates (OPs), chlorpyrifos and diazinon, with an insect growth regulator (IGR), pyriproxyfen. Following laboratory studies for 12 months, field trials were conducted in six experimental huts, randomly allocated to one or two layers of insecticide at 167 g/m² painted on wall surfaces with a brush, or an untreated control. When two layers were applied, the first layer was diluted in 20% water. Mortality was still 90–100% against pyrethroid-resistant mosquito populations 6 months after treatment and where two layers had been applied, mortality was still 90–93% against *An. gambiae* and 55% against *Culex quinquefasciatus* at 9 months.

Innovative application technique

Devine et al. (2009) has shown that a larvicide, such as pyriproxyfen, can be distributed to oviposition sites by using adult mosquitoes. Trials showed that deposits of the growth hormone at *Ae. aegypti* resting sites were picked up by adult mosquitoes and deposited on water surfaces during oviposition. Pyriproxyfen had no effect on the adult, but was highly effective at very low doses on young larvae when moulting to the next instar.

Genetically modified mosquitoes

The idea of using sterile mosquitoes to achieve control is not new. However, the earlier attempts were stopped by ill-informed press reports and political campaigns (Curtis and Townson, 1998). The sterile insect technique (SIT) was pioneered by Knipling (1955) and has been used successfully to control a number of agricultural insect pests, most notably the screw worm. Large numbers of the insects are reared and the males sterilised before release, so that by mating with wild females, the wild population is reduced in the subsequent generation. Continuing to release over a sufficient period will result in a collapse of the target population and elimination if there is no re-invasion into the treated area.

In contrast to applying broad spectrum insecticides, releasing sterile insects as a strategy has a number of advantages, including being species-specific, thus avoiding effects on non-target species. It utilises on the natural biology of the species as the male insects seek females to mate and becomes more efficient as the wild population decreases. The technique is thus suitable when the aim is to eradicate the vector, but there are technological problems to be overcome (Coleman and Alphey, 2004).

Releasing male mosquitoes, which do not bite humans, would be acceptable but this means that the males must be separated from the females during mass rearing. There is a technique for separating male and female pupae of *Culex* and *Aedes* mosquitoes based on their size, but this was not effective for Anopheline vectors of malaria. Another technique developed using *Drosophila*, known as the Release of Insects with a Dominant Lethal (RIDL) gene (Alphey and Andreasen, 2002), may be a way forward but further research is needed. The technique does not require sterilisation of the insects and relies on producing only male insects, which when mating in the wild produce only viable males. More recently, Fu et al. (2010) reported on using either two separate transgenes or a single transgene, based on the use of a female-specific indirect flight muscle promoter from the *Ae. aegypti* Actin-4 gene to engineer a repressible flightless female-specific phenotype. With this development of genetic sexing, male-only releases can be made without the need for irradiation and enable the release of eggs instead of adults to facilitate area-wide dengue control. However, even if successful, if used against one malaria vector species, another malaria vector could quickly replace it (Riehle et al., 2003). The use of fluorescent marker to show a genetic modification is shown in Figures 8.2 and 8.3.

Another dominant gene technique uses Tetracycline-repressible lethality in LA513A, so that when the mosquito population is mass reared in the laboratory, larvae are fed tetracycline and survive, but when these are released in the field, the larvae do not have access to the tetracycline and die. There were plans to evaluate this technique in Malaysia with *Ae. aegypti* (Phuc et al., 2007) and a small-scale trial was conducted successfully in the Cayman Islands with the release of 3–6-day-old adult RIDL males, strain OX513A, three times a week. Trapping adults showed a significant change in the male/female ratio and the proportion of ovitraps containing one or more eggs declined significantly compared with a similar control area (Oxitec, 2010).

A different approach is to use genetic techniques with artificial constructs to encode peptides that reduce the ability of mosquitoes to transmit malaria or dengue virus (Ito, 2002; Sinkins and Gould, 2006). Recently, a malaria-proof mosquito has been engineered, which introduced a gene that

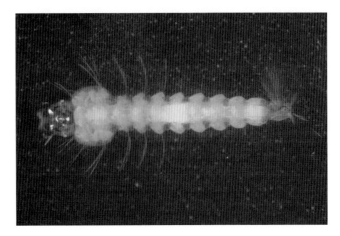

Figure 8.2 Genetically modified mosquito – larva showing fluorescence marker indicating that the genetic modification was present (photo: Michael Reihle).

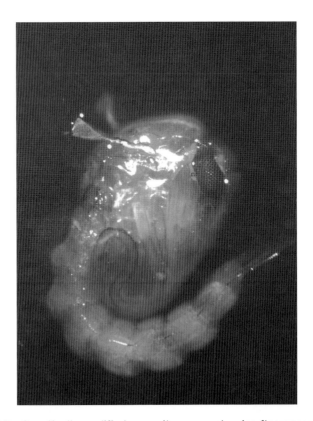

Figure 8.3 Genetically modified mosquito – pupa showing fluorescence marker indicating that the genetic modification was present (photo: Michael Reihle).

affected the insect's gut so that the malaria parasite could not develop (Corby-Harris et al., 2010).

While tools have been developed to genetically modify mosquitoes, there remains an extensive gap between laboratory studies and the ability to transfer this technology to the field. Different species of *Anopheles* mosquitoes transmit malaria, thus in a survey in one rural area of Benin, seven different species were collected (Djenontin et al., 2010), so the effect of controlling one species, such as *An gambiae*, may be an increase of another species, such as *An. funestus*. One hurdle to overcome is the ability of the transformed mosquitoes to compete with the wild type. Large numbers of the GM mosquitoes would need to be mass produced to treat an effective area continually, so that sufficient mate with invaders from outside the area and keep the population at a low level. If the aim is eradication, then large-scale trials would need to be conducted on a very large area-wide basis and re-invasion from the periphery would also need to be controlled. Some other technique may have to be used prior to releases, to reduce the population and permit release of fewer GM mosquitoes. If chemical control is needed, perhaps an area-wide space treatment (fog) with a non-persistent insecticide might be one possibility.

Attractants

Traps have been used primarily to monitor populations of vectors, thus mosquito populations have been sampled using various types of traps, such as the Center for Disease Control and Prevention in the USA (CDC) Miniature trap, preferably used with an attractant, for example, dry ice to emit CO_2 or octenol and by using gravid traps. Gravid traps have been considered to be more useful in detecting mosquitoes early in the season (White et al., 2009). A new trap, the Magnet Pro™ trap, has also been useful in sampling mosquito populations (Korgaonkar et al., 2008) and used in conjunction with the Elisa test to determine the proportion of infective mosquitoes.

However, there have been studies to investigate whether using traps would alleviate the nuisance caused by mosquitoes. Qualls and Mullen (2007) showed that by using 1-octen-3-ol significantly enhanced collection of mosquitoes in a Magnet Pro™ trap, with a 3-fold increase in the collection of *Anopheles punctipennis* and an 18-fold increase of *Aedes vexans* at one site. However, Kline (2006, 2007) reported that using these traps around the perimeter of a residential area did not reduce the level of mosquito nuisance, although there was a significant reduction of the black salt-marsh mosquito, *Ochlerotatus taeniorhynchus*, when the traps were placed along a nature trail on an island. Henderson et al. (2006) also failed to show any significant suppression of mosquito activity in tests in Winnepeg, Canada. Thus, their use may only have a limited place to keep mosquito populations low in certain situations.

Recent studies have examined the attractiveness of a synthetic blend of CO_2, ammonia and carboxylic acids to mosquitoes in experimental huts in Africa. The experimental blend was 3–5 times more attractive than humans, when the odour blend was in a different hut to the humans, but it was equally or less effective when the humans were in the same hut (Okuma et al., 2010). Further studies may lead to a more effective way of attracting mosquitoes.

Attractant toxic baits have shown significant reductions of *Anopheles* mosquitoes in a trial in Mali, which was referred to in Chapter 6 under Barrier treatments.

Greater success at using traps as a means of controlling the tsetse fly in Africa, was achieved where the attractant octenol was used with pyrethroid insecticide treated black cloth screens (Vale et al., 1988). Such traps have since been widely used, where access is possible to re-treat the screens and replenish the attractant.

Urbanisation

People throughout the world have been attracted to congregate in towns. expecting a better life-style with more amenities. Only 29% of the world's population were living in urban areas in 1950. However, the trend has accelerated, with over 70% in developed countries now in urban areas, while in the developing countries it is expected to be over 50% by 2025. Sadly, the rapid population movement in the poorer countries towards the cities has led to poor levels of hygiene, and increasing urban poverty. The urban environment in many developing countries has deteriorated, with densely packed housing in shanty towns or slums and inadequate drinking-water supplies, garbage collection services, and surface-water drainage. Thus, the habitat has become favourable for the vectors of disease, such as malaria and dengue (Figure 8.4).

Back in 1992, it was suggested that the problems in controlling vector borne diseases or possibly eliminating the vectors required decision-makers and urban planners to avoid the concept of 'blanket' applications of pesticides and adopt integrated approaches (Knudsen and Slooff, 1992). It has been argued that sound environmental management practices and community education and participation are essential parts of the control of vectors.

Keiser et al. (2004) evaluated the impact of urbanisation in Africa on malaria and concluded with an estimate of annual incidence of 24.8–103.2 million cases of clinical malaria attacks would occur among those living in urban areas in sub-Saharan Africa. They recommended that the promotion of insecticide-treated bed nets for the rapidly growing numbers of the urban poor must be improved, alongside well-tailored and integrated malaria control strategies. Also, as poor housing and lack of sanitation and drainage of surface water can increase vector breeding and human vector contact in urban areas, they proposed environmental management and larviciding within well-defined sites as a main feature for such an integrated control approach.

In a study comparing an urban area and rural village in Cameroon, seven species of *Anopheles* were found in the village compared to five species in the urban area, with the annual entomological inoculation rate estimated at 129 infective bites/person/year in the urban area and 322 in the village, where

Figure 8.4 Urbanisation. A housing area in part of the Cameroon capital, Yaounde.

94% of the transmission was due to *A. moucheti* in contrast to *An. gambiae* in the urban area (Antonio-Nkondjio et al., 2005). The study in Cameroon only looked at two places, but Robert et al. (2003) reported that overall anopheline densities were also significantly higher in the rural than in the semi-urban environment. Their results showed mean annual entomologic inoculation rates (EIR) of 7.1 in the city centres, 45.8 in peri-urban areas, and 167.7 in rural areas, but with considerable variation between cities and within districts in the same city. At a meeting in Pretoria in 2004, The Pretoria Statement on Urban Malaria stated that urban malaria in sub-Saharan Africa is a major health problem and is likely to increase in importance unless addressed, and that with a range from high-density neighbourhoods to peri-urban agricultural zones, 'urban malaria is amenable to cost-effective prevention and control by tailoring existing tools for the diagnosis and treatment of infection, and for vector control.' (Donnelly et al., 2005). By managing malaria in urban areas, there is an immediate benefit to businesses as well as to the communities. In Mozambique, an IRS programme was shown to reduce suffering due to malaria at a lower cost per person per year in a peri-urban ($2.41), compared to a rural situation ($3.86) (Conteh et al., 2004), but to sustain this does require financial support, political will, collaborative management and training and community involvement. Similarly in Zambia, one round of IRS in mining towns had an immediate effect of reducing malaria (Sharp et al., 2002).

While malaria control may be easier to implement in an urban environment with the availability of medicines from accredited pharmacies in the private sector and the more readily availability of rapid diagnostic tests, environmental management and larviciding to reduce larval breeding remains a crucial part of integrated vector management together with promotion of insecticide-treated bed nets and IRS. For vector control, the urban area with a high population requiring treatment is limited and more easily defined compared to extensive rural areas, but the problems are different to those confronted with malaria control in rural areas. Protection of those in rural areas remains crucially important if a country is to improve agricultural production and avoid too many people seeking to live in towns and cities. While use of bed nets by children may reduce mortality, the key to reducing poverty must lie in creating more employment and for agriculture this will be in a rural environment.

In some situations, a community approach has been successful, thus in Santiago de Cuba, where *Ae. aegypti* is the key vector, stakeholders within a neighbourhood were asked to form a task force to assess the perceived needs and develop action plans to reduce environmental risks through social communication strategies and inter-sectoral local government activities (Toledo et al., 2007). With this activity, the number of uncovered water storage containers decreased from 49.3 to 2.6% between 2000 and 2002. People did not like the taste of water treated with temephos, but education made them aware of the importance of using a larvicide and stopped them removing the insecticide. These activities reduced by 75% the number of water containers tested positively with larvae. The frequency of garbage collection was increased from two to four times a month, and repairs to the water supply network resulted in improving the delivery of water from every 15 days to 7 days.

These activities as part of integrated vector management show that vector populations can be significantly reduced and will continue to be crucial, even when new technology such as the release of GM mosquitoes is approved and is shown to be effective, as it will enable the new technology to operate on a lower vector population.

Economics

African economies have depended on export of minerals and basic agricultural commodities, without value added. Over the last five decades, many sub-Saharan countries have suffered conflicts, especially in the countries that export highly sought after or high value items, notably oil and diamonds. In consequence, there has been relatively little industrial development in Africa compared to other continents, so survival of governments has depended to a large extent on aid, both in the pre- and post-independence eras. Most of the people in sub-Saharan Africa countries, with a few exceptions, have remained poor, whether in rural or urban areas. South Africa is clearly different, with industrial and mining developments and. to a lesser extent, this was also true in Zimbabwe, where local industrial development had enabled the country to have a more balanced economy, prior to the political problems concerning ownership of land. The lack of income generated locally has undoubtedly had an impact on the extent of finance for health care and in particular vector control, which is not a mainstream part of medical education.

Thus, African countries will depend on outside aid to help overcome the debilitating impact of vector borne diseases, together with other changes to alleviate poverty if people are to enjoy better lives. The cost of even the simplest vector control intervention is beyond the income of most people in sub-Saharan Africa, yet there is a need to integrate different interventions on an area-wide scale to have a real and hopefully lasting impact on the vectors. To achieve this, major investment and commitment by governments over at least a decade will be required. The WHO, through the Roll Back Malaria programme and Global Fund assisted by other major funding agencies, have made significant progress in stimulating actions to reduce the burden of malaria. Vector control will continue to be a key part of this to minimise the need to use drugs such as artemisinin combination therapy (ACT), and confine its use to those who have confirmed malaria by modern diagnostic tools. Progress with reducing malaria and improvement in health care will hopefully also lead to a greater control of other tropical diseases and help people escape from poverty. However, for this to be accomplished, there is a need for commitment of governments combined with a concurrent need to help establish improved farming practices and other employment, with better housing to reduce poverty. Initially success can probably be achieved in areas with malaria confined to a part of the year, as in southern Africa, but people living in the humid tropics can also benefit by concerted efforts through integrated vector management programmes, provided there is community involvement, commitment by governments and adequate trained personnel for all aspects of a control and monitoring programme.

Conclusion

In looking ahead, it is good that there has been excellent funding for research seeking new technologies, but sadly there has been virtually no support for operational research to improve existing technology (Roberts et al., 2000) or adequate support for training personnel and the infrastructure needed to maintain vector control over a sufficient period to have a sustained impact. Good integrated vector management adapted to suit local circumstances of vector biology can bring down vector populations and reduce disease impact, but at present there is the risk of a subsequent epidemic as soon as efforts are relaxed due to lack of funds or other reasons.

Certainly reliance on one control technique will fail, but even if treated bed nets, IRS and larviciding areas that cannot be drained are promoted with adequate monitoring alongside the control measures, ultimately improved housing will also be needed. Preventing mosquitoes entering houses using screens has proved to be both an effective and long-lasting way of reducing the incidence of malaria. Development of a vaccine against malaria has been in progress for many years and at least one is now in Stage III trials in Africa.

Hopefully such a vaccine can be developed to protect young children and others exposed to malaria, but vector control will still have a role to reduce the vectors of other diseases and reduce the impact of nuisance insects.

In most countries with major vector-borne diseases, training local staff to develop and adapt local campaigns integrated with similar activity throughout a country is crucial to maintain vector populations at a low level.

It is hoped that this book will provide some help in deciding on actions needed to improve the implementation of existing technology, as well as development of improved control of the vectors.

References

Alphey, L. and Andreasen, M. (2002) Dominant lethality and insect population control. *Molecular and Biochemical Parasitology* **121**: 173-8.

Amelotti, I., Catalá, S. S. and Gorla, D. E. (2009) Experimental evaluation of insecticidal paints against *Triatoma infestans* (Hemiptera: Reduviidae), under natural climatic conditions. *Parasites and Vectors* **2**: 30-6.

Antonio-Nkondjioa, C., Simarda, F., Awono-Ambenea, P., et al. (2005) Malaria vectors and urbanization in the equatorial forest region of south Cameroon. *Transactions of the Royal Society of Tropical Medicine and Hygiene* **99**: 347-54.

Assidi, A. N., N'Guessan, R. N., Hutchinson, R. A., Traore-Lamizana, M., Carnevale, P. and Curtis, C. F. (2004) Experimental hut comparisons of nets treated with a carbamate or pyrethroid insecticides, washed or unwashed, against pyrethroid resistant mosquitoes. *Medical and Veterinary Entomology* **18**: 134-40.

Bell, A. S., Blanford, S., Jenkins, N., Thomas, M. B. and Read, A. F. (2009) Real-time quantitative PCR for analysis of candidate fungal biopesticides against malaria: technique validation and first applications. *Journal of Invertebrate Pathology* **100**: 160-8.

Blanford, S., Chan, B. H. K., Jenkins, N., et al. (2005) Fungal pathogen reduces potential for malaria transmission. *Science* **308**: 1638-41.

Chadd, E. M. (1990) Use of an electrostatic sprayer for control of anopheline mosquitoes. *Medical and Veterinary Entomology* **4**: 97-104.

Coleman. P. G. and Alphey, L. (2004) Genetic control of vector populations: an imminent prospect. *Tropical Medicine and International Health* **9**: 433-7.

Conteh, L., Sharp, B. L., Streat, E., Barreto, A. and Konar, S. (2004) The cost and cost-effectiveness of malaria vector control by residual insecticide house-spraying in southern Mozambique: a rural and urban analysis. *Tropical Medicine and International Health* **9**: 125-32.

Corby-Harris, V., Drexler, A., Watkins de Jong, L., et al. (2010) Activation of Akt signaling reduces the prevalence and intensity of malaria parasite infection and lifespan in *Anopheles stephensi* mosquitoes. *PLoS Pathology* **6**: e1001003.

Curtis, C. F. and Townson, H. (1998) Malaria: existing methods of vector control and molecular entomology. *British Medical Bulletin* **54**: 311-25.

Curtis, C. F. (2010) Current prospects for the control of the vectors of malaria and filiariasis. In: Atkinson, P. W. (ed.), *Vector Biology, Ecology and Control*. Springer. pp. 179-90.

Darbro, J. M. and Thomas, M. B. (2009) Spore persistence and likelihood of aero-allergenicity of entomopathogenic fungi used for mosquito control. *Journal of the American Mosquito Control Association* **80**: 992-7.

Darriet, F., N'Guessan, R., Koffi, A., Doannio, I. M. C., Chandre, F. and Carnevale, P. (2000) Impact de la resistance aux pyrethinroides sur l'efficacite desmo-usiquaires impregnees dans la prevention du paludisme: resultants des essays en cases experimentales aves deltamethrine SC. *Bulletin of the Exotic Pathology Society* **95**: 131-4.

Devine, G. J., Zamora Pere, E., Stancile J. D., Clarke, S. J. and Morrison, A. C. (2009) Using adult mosquitoes to transfer insecticide to *Aedes aegypti* larval habitats. *Proceedings of the National Academy of Sciences* **106**: 11,530-4.

Djenontin, A., Bio-Bangana, S., Moiroux, N., et al. (2010) *Culicidae* diversity, malaria transmission and insecticide resistance alleles in malaria vectors in Ouidah-Kpomasse-Tori district from Benin (West Africa): a pre-intervention study. *Parasites and Vectors* **3**: 83.

Donnelly, M. J., McCall, P. J., Lengeler, C., et al. (2005) Malaria and urbanization in sub-Saharan Africa. *Malaria Journal* **4**: 12.

Farenhorst, M., Farina, D., Scholte, E.-J., et al. (2008) African water storage pots for the delivery of the entomopathogenic fungus *Metarhizium anisopliae* to the malaria vectors *Anopheles gambiae* s.s. and *Anopheles funestus*. *Journal of the American Mosquito Control Association* **78**: 910-16.

Farenhorst, M., Mouatchob, J. C., Kikankieb, C. K., et al. (2009) Fungal infection counters insecticide resistance in African malaria mosquitoes. *Proceedings of the National Academy of Sciences of the United States of America* **106**: 17,443-7.

Farenhorst, M., Knols, B. G. J., Thomas, M. B., et al. (2010) Synergy in efficacy of fungal entomopathogens and Permethrin against West African insecticide-resistant *Anopheles gambiae* mosquitoes. *PLoS ONE* **5**: e12081.

Fu, G., Lees, R. S., Nimmo, D., et al. (2010) Female-specific flightless phenotype for mosquito control. *Proceedings of the National Academy of Sciences* **107**: 4550-4.

Hemingway, J., Beaty, B. J., Rowlands, R., Scott, T. W. and Sharps, B. L. (2006) The Innovative Vector Control Consortium: improved control of mosquito-borne diseases. *Trends in Parasitology* **22**: 308-9.

Henderson, J. P., Westwood, R. and Galloway, T. (2006) An assessment of the effectiveness of the mosquito Magnet Pro™ model for suppression of nuisance mosquitoes. *Journal of the American Mosquito Control Association* **22**: 401-7.

Ito, J., Ghosh, A., Moreira, L. A., Wimmer, E. A. and Jacobs-Lorena, M. (2002) Transgenic anopheline mosquitoes impaired in transmission of a malaria parasite. *Nature* **417**: 452-5.

Keiser, J., Utzinger, J., De Castro, M. C., Smith, T. A., Tanner, M. and Singer, B. H. (2004) Urbanization in sub-Saharan Africa and implication for malaria control. *Journal of the American Mosquito Control Association* **71(suppl 2)**: 118-27.

Kline, D. L. (2006) Traps and trapping techniques for adult mosquito control. *Journal of the American Mosquito Control Association* **22**: 490-6.

Kline, D. L. (2007) Semiochemicals, traps/targets and mass trapping technology for mosquito management. *AMCA Bulletin* No. 7. **23(suppl 2)**: 241-51.

Korgaonkar, N., Kumar, A., Yadav, R. S., Kabadi, D. and Dash, A. (2008) Sampling of adult mosquito vectors with mosquito Magnet Pro™ in Panaji, Goa, India. *Journal of the American Mosquito Control Association* **24**: 604-7.

Knipling, E. (1955) Possibilities of insect control or eradication through the use of sexually sterile males. *Journal of Economic Entomology* **48**: 459-66.

Knudsen, A. B. and Slooff, R. (1992) Vector-borne disease problems in rapid urbanization: new approaches to vector control. *Bulletin of the World Health Organization* **70**: 1-6.

Malamud-Roam, K., Cope, S. E. and Stickman, D. (2010) IR-4: the new partner in the search for public health pesticides. *Wing Beats* **21(3)**: 13-15.

Mosqueira, B., Duchon, S., Chandre, F., Hougard, J.-M., Carnevale, P. and Mas-Coma, S. (2010a) Efficacy of an insecticide paint against insecticide-susceptible and resistant mosquitoes. Part 1: Laboratory evaluation. *Malaria Journal* **9**: 340.

Mosqueira, B., Chabi, J., Chandre, F., et al. (2010b) Efficacy of an insecticide paint against malaria vectors and nuisance in West Africa. Part 2: Field evaluation. *Malaria Journal* **9**: 341.

N'Guessan, R., Corbel, V., Bonnet, J., et al. (2007a) Evaluation of indoxacarb, an oxadiazine insecticide for the control of pyrethroid-resistant *Anopheles gambiae* (Diptera: Culicidae). *Journal of Medical Entomology* **44**: 270-6.

N'Guessan, R., Corbel, V., Akogbeto, M. and Rowland, M. (2007b) Reduced efficacy of insecticide treated nets and indoor residual spraying for malaria control in pyrethroid resistant area. *Benin Emerging Infectious Diseases* **1**: 199-206.

N'Guessan, R., Boko, P., Odjo, A., Knols, B., Akogbeto, M. and Rowland, M. (2009) Control of pyrethroid-resistant *Anopheles gambiae* and *Culex quinquefasciatus* mosquitoes with chlorfenapyr in Benin. *Tropical Medicine and International Health* **14**: 389-95.

Okumu, F. O., Killeen, G. F., Ogoma, S., et al. (2010) Development and field evaluation of a synthetic mosquito lure that is more attractive than humans. *PLoS One* **5(1)**: p. e8951.

Oxborough, R. M., Kitau, J., Matowo, J., et al. (2010) Evaluation of indoor residual spraying with the pyrrole insecticide chlorfenapyr against pyrethroid-susceptible *Anopheles arabiensis* and pyrethroid-resistant *Culex quinquefasciatus* mosquitoes. *Transactions of the Royal Society of Tropical Medicine and Hygiene* **104**: 639-45.

Oxitec (2010) Suppression of a field population of *Aedes aegypti* using the RIDL system. Abstract 1020: Florida Mosquito Control Association meeting November 2010.

Paul, A., Harrington, L. C. and Scott, J. G. (2006) Evaluation of novel insecticides for control of dengue vector *Aedes aegtpti* (Diptera: Culicidae). *Journal of Medical Entomology* **43**: 55-60.

Penilla, R. P., Rodríguez, A. D., Hemingway, J., Trejo, A., López, A. D. and Rodríguez, M. H. (2007) Cytochrome P450-based resistance mechanism and pyrethroid resistance in the field *Anopheles albimanus* resistance management trial. *Pesticide Biochemistry and Physiology* **89**: 111-17.

Phuc, H. K., Andreasen, M. H., Burton, R. S., et al. (2007) Late-acting dominant lethal genetic systems and mosquito control. *BMC Biology* **5**: 11.

Qualls, W. A. and Mullen, G. R. (2007) Evaluation of the Mosquito Magnet Pro™ trap with and without 1-octen-3-ol for collecting *Aedes albopictus* and other urban mosquitoes. *Journal of the American Mosquito Control Association* **23**: 131-6.

Rajatileka, S., Burhani, J. and Ranson, H. (2011) Mosquito age and susceptibility to insecticides. *Transactions Royal Society Tropical Medicine and Hygiene* **105**: 247-53.

Ranson, H., N'Guessan, R., Lines, J., Moiroux, N., Nkuni, Z. and Corbel, V. (2011) Pyrethroid resistance in African anopheline mosquitoes: what are the implications for malaria control? *Trends in Parasitology* **27**: 91-8.

Read, A. F., Lynch, P. A. and Thomas, M. B. (2009) How to make evolution-proof insecticides for malaria control. *PLoS Biol* **7**: e1000058.

Riehle, M. A., Srinivasan, P., Moreira, C. K. and Jacobs-Lorena, M. (2003) Towards genetic manipulation of wild mosquito populations to combat malaria: advances and challenges. *Journal of Experimental Biology* **206**: 3809-16.

Robert, V., Macintyre, K., Keating, J., et al. (2003) Malaria transmission in urban sub-Saharan Africa. *Journal of the American Mosquito Control Association* **68**: 169-76.

Roberts, D. R., Manguin, S. and Mouchet, J. (2000) DDT house spraying and re-emerging malaria. *Lancet* **356**: 330-2.

Scholte, E.-J., Knols, B. G. J., Samson, R. A. and Takken, W. (2004) Entomopathogenic fungi for mosquito control: a review. *Journal of Insect Science* **4**: 19.

Scholte, E.-J., Knols, B. G. J. and Takken, W. (2006) Infection of the malaria mosquito *Anopheles gambiae* with the entomopathogenic fungus *Metarhizium anisopliae* reduces blood feeding and fecundity. *Journal of Invertebrate Pathology* **91**: 43-9.

Sharp, B., van Wyk, P., Sikasote, J. B., Banda, P. and Kleinschmidt, I. (2002) Malaria control by residual insecticide spraying in Chingola and Chililabombwe, Copperbelt Province, Zambia. *Tropical Medicine and International Health* **7**: 732-6.

Sinkins, S. P. and Gould, F. (2006) Gene drive systems for insect disease vectors. *Nature Reviews Genetics* **7**: 427-35.

Skovmand, O. (2010) Insecticidal bed nets for the fight against malaria - present time and near future. *The Open Biology Journal* **3**: 92-6.

Thomas, M. B. and Read, A. F. (2007) Can fungal biopesticides control malaria? *Nature Reviews Microbiology* **5**: 377-83.

Toledo, M. E., Vanlerbergheb, V., Balya, A., et al. (2007) Towards active community participation in dengue vector control: results from action research in Santiago de Cuba, Cuba. *Transactions of the Royal Society of Tropical Medicine and Hygiene* **101**: 56-63.

Turley, A. P., Moreira, L. A., O'Neill, S. L. and McGraw, E. A. (2009) *Wolbachia* infection reduces blood-feeding success in the Dengue fever mosquito, *Aedes aegypti*. *PLoS Neglected Tropical Diseases* **3(9)**: e516. doi:10.1371/journal.pntd.0000516

Vale, G. A., Lovemore, D. F., Flint, S. and Cockbill, G. F. (1988) Odour-baited targets to control tsetse flies, *Glossina* spp. (Diptera: Glossinidae) in Zimbabwe. *Bulletin of Entomological Research* **78**: 31-49.

White, S. L., Ward, M. P., Budke, C. M., Cyr, T. and Bueno, R. (2009) A comparison of gravid and underhouse CO_2-baited CDC light traps for mosquito species of public health importance in Houston, Texas. *Journal of Medical Entomology* **46**: 1494-7.

Zaim, M. and Guillet, P. (2002) Alternative insecticides: an urgent need. *Trends in Parasitology* **18**: 161-3.

Appendix A
Calibration

It is important to check the flow rate from spray nozzles and calculate the amount of spray that will be applied to ensure that the correct dose of insecticide is used.

On small equipment, it is necessary to measure the flow rate from the nozzle at the operational pressure, the width of the swath being treated and the speed at which the nozzle is moved across a surface.

Measure: Output in litres per minute (l/min). Swath in metres (m) and travel speed in metres per minute (m/min). For indoor residual spraying, the speed across a wall may be measured initially as metres per second and then multiplied by 60 to convert to per minute.

Require: Sprayer complete with nozzle combined with the control flow valve*, water, stopwatch, bucket/measuring cylinder, measuring tape.

Calculation: *For indoor residual spraying*

$$\frac{\text{Output (l/min)}}{\text{Swath (m)} \times \text{Speed (m/min)}} = \text{litres per m}^2$$

For larger areas of spraying

$$\frac{\text{Output (l/min)}}{\text{Swath (m)} \times \text{Speed (m/min)}} \times 10,000 = \text{litres per hectare}$$

For space treatments with vehicle or aerial equipment, instead of 'swath' when using hydraulic nozzles, the swath is replaced by the **Track spacing** required for operating the sprayer.

$$\frac{\text{Output (l/min)}}{\text{Tracking Spacing (m)} \times \text{Speed (m/min)}} \times 10,000 = \text{litres per hectare}$$

* See Chapter 2

Integrated Vector Management: Controlling Vectors of Malaria and Other Insect Vector Borne Diseases, First Edition. Graham Matthews.
© 2011 John Wiley & Sons, Ltd. Published 2011 by John Wiley & Sons, Ltd.

Appendix B
Conversion Tables

	A	B	A→B	B→A
Weight	oz	g	× 28.35	× 0.0353
	lb	kg	× 0.454	× 2.205
	cwt	kg	× 50.8	× 0.0197
	ton (long)	kg	× 1016	× 0.000984
	ton (short)	ton (long)	× 0.893	× 1.12
Surface area	in^2	cm^2	× 6.45	× 0.155
	ft^2	m^2	× 0.093	× 10.764
	yd^2	m^2	× 0.836	× 1.196
	yd^2	acre	× .000207	× 4840
	acre	ha	× 0.405	× 2.471
Length	μm	mm	× 0.001	× 1000
	in	cm	× 2.54	× 0.394
	ft	m	× 0.305	× 3.281
	yd	m	× 0.914	× 1.094
	mile	km	× 1.609	× 0.621
Velocity	ft/s	m/s	× 0.305	× 3.281
	ft/min	m/s	× 0.00508	× 197.0
	mile/h	km/h	× 1.609	× 0.621
	mile/h	ft/min	× 88.0	× 0.0113
	knot	ft/s	× 1.689	× 0.59
	m/s	km/h	× 3.61	× 0.277
	cm/s	km/h	× 0.036	× 27.78
Quantities/area	lb/acre	kg/ha	× 1.12	× 0.894
	lb/acre	mg/ft^2	× 10.4	× 0.09615
	kg/ha	mg/m^2	× 100	× 0.01
	mg/ft^2	mg/m^2	× 10.794	× 0.093
	oz/yd^2	cwt/acre	× 2.7	× 0.37
	gal (Imp.)/acre	litre/ha	× 11.23	× 0.089
	gal (USA)/acre	litre/ha	× 9.346	× 0.107

(Continued)

Integrated Vector Management: Controlling Vectors of Malaria and Other Insect Vector Borne Diseases, First Edition. Graham Matthews.
© 2011 John Wiley & Sons, Ltd. Published 2011 by John Wiley & Sons, Ltd.

(*Continued*)

	A	B	A→B	B→A
	fl oz (Imp.)/ acre	ml/ha	× 70.05	× 0.0143
	fl oz (USA)/ acre	ml/ha	× 73.14	× 0.0137
	oz/acre	g/ha	× 70.05	× 0.0143
	oz/acre	kg/ha	× 0.07	× 14.27
Dilutions	fl oz/100 gal (Imp.)	ml/100 litres	× 6.25	× 0.16
	pint/100 gal (Imp.)	ml/100 litres	× 125	× 0.008
	oz/gal (Imp.)	g/litre	× 6.24	× 0.16
	oz/gal (USA)	g/litre	× 7.49	× 0.134
	lb/100 gal (Imp.)	kg/100 litres	× 0.0998	× 10.02
Density of water	gal (Imp.)	lb	× 10	× 0.1
	gal (USA)	lb	× 8.32	× 0.12
	lb	ft^3	× 0.016	× 62.37
	litre	kg	× 1	× 1
	ml	g	× 1	× 1
	lb/gal (Imp.)	g/ml	× 0.0997	× 10.03
	lb/gal (USA)	g/ml	× 0.1198	× 8.34
	lb/ft^3	kg/m^3	× 16.1	× 0.0624
Volume	in^3	ft^3	× 0.000579	× 1728
	ft^3	yd^3	× 0.037	× 27
	yd^3	m^3	× 0.764	× 1.308
	fl oz (Imp.)	ml	× 28.35	× 0.0352
	fl oz (USA)	ml	× 29.6	× 0.0338
	gal (Imp.)	gal (USA)	× 1.20	× 0.833
	gal (Imp.)	litre	× 4.55	× 0.22
	gal (USA)	litre	× 3.785	× 0.264
	cm^3	m^3	× 10^{-6}	× 10^6
	cm^3	μm	× 10^{12}	× 10^{-12}
Pressure	lb/in^2	kg/cm^2	× 0.0703	× 14.22
	lb/in^2	bar	× 0.0689	× 14.504
	bar	kPa	× 100	× 0.01
	lb/in^2	kPa	× 6.89	× 0.145
	kN/m^2	kPa	× 1	× 1
	N/m^2	kPa	× 0.001	× 1000
	lb/m^2	atm	× 0.068	× 14.696
Power	hp	kW	× 0.7457	× 1.341
Temperature	°C	°F	$\dfrac{9}{5}$ °C + 32	$\dfrac{5}{9}$ (°F − 32)

Index

acaricide, 41
acute oral toxicity, 17
aerial granule application,
144-8
aerial larvicide application,
140-2, 150-2
aerial space treatments, 86-92
aerial spraying, 4, 11, 66, 68, 86-90,
97, 132, 140, 142
aerosol, 20, 63-5, 68, 90, 96, 191,
192, 196
African Programme for
Onchocerciasis Control
(APOC), 132
airblower, 80
aircraft, 63, 87, 152, 171, 196
'airport malaria', 1
American Forces Pest Management
Board (AFPMB), 203
ancilliary equipment, 52
AngloGold Ashanti Mining
Company, 57
anthranilic diamide, 204
artemisinin combination therapy
(ACT), 46, 164, 180, 217
attractants, 213-14

Bacillus sphericus, 17, 18,
138, 164

Bacillus thuringiensis, 202
Bacillus thuringiensis israelensis
(Bti), 17, 18, 76, 132, 137, 138,
140, 142, 144, 150, 151,
164, 207
baits, 4, 192, 195-6
barrier treatment, 4, 79, 164, 167,
169, 175
battle dress uniforms (BDUs), 123
bazaar fly, 13
Beauvaria bassiana, 207
bed nets, 4, 5, 9, 13, 26, 44, 57,
105-22, 137, 164, 166, 173, 177,
182-4, 204, 206, 217
coating technology, 108
disposal, 122
distribution 117
incorporation technology, 108
mesh size, 109
operational use, 120
social marketing, 118
washing, 116-17
biconical trap, 10
bioassays, 40, 53-5, 58, 116, 186
biopesticide *see* insecticides
birds, 56, 137
biting rates, 26
bitter agent, 192, 197
boat, 4, 25, 150

Integrated Vector Management: Controlling Vectors of Malaria and Other Insect Vector Borne Diseases, First Edition. Graham Matthews.
© 2011 John Wiley & Sons, Ltd. Published 2011 by John Wiley & Sons, Ltd.

Forest Entomology
A Global Perspective
by
William Ciesla

Insects are the most abundant and diverse organisms that inhabit our planet and are found in all the world's forest ecosystems. *Forest Entomology: A Global Perspective* examines forest insects in a global context and reviews their dynamics, interactions with humans and methods for monitoring and management of species that damage forests. Also provided are 235 profiles of forest insects, worldwide. A series of tables provides summaries of the distribution and hosts of many more species. Included are those that damage forests, others that are simply curiosities and some that are beneficial. This book is designed as a reference for students, practicing foresters and forest health specialists, especially for those who work internationally or are concerned with species that have the potential to expand their ranges via international trade, travel or environmental changes.

Go online to order your copy today: www.wiley.com/go/entomology

Online Version also available at: www.wileyonlinelibrary.com

Print ISBN: 978-1-4443-3314-5